Beyond Curie
Four women in physics and their remarkable discoveries, 1903 to 1963

Beyond Curie
Four women in physics and their remarkable discoveries, 1903 to 1963

Scott Calvin
Lehman College of the City University of New York

Morgan & Claypool Publishers

Copyright © 2017 Morgan & Claypool Publishers

All rights reserved. No part of this publication may be reproduced, stored in a retrieval system or transmitted in any form or by any means, electronic, mechanical, photocopying, recording or otherwise, without the prior permission of the publisher, or as expressly permitted by law or under terms agreed with the appropriate rights organization. Multiple copying is permitted in accordance with the terms of licences issued by the Copyright Licensing Agency, the Copyright Clearance Centre and other reproduction rights organisations.

Rights & Permissions
To obtain permission to re-use copyrighted material from Morgan & Claypool Publishers, please contact info@morganclaypool.com.

ISBN   978-1-6817-4645-6 (ebook)
ISBN   978-1-6817-4644-9 (print)
ISBN   978-1-6817-4647-0 (mobi)

DOI   10.1088/978-1-68170-4645-6

Version: 20170701

IOP Concise Physics
ISSN 2053-2571 (online)
ISSN 2054-7307 (print)

A Morgan & Claypool publication as part of IOP Concise Physics
Published by Morgan & Claypool Publishers, 40 Oak Drive, San Rafael, CA, 94903 USA

IOP Publishing, Temple Circus, Temple Way, Bristol BS1 6HG, UK

For Erin Eisenbarth

Contents

Acknowledgments		ix
Author biography		x
1	**Introduction**	**1-1**
1.1	Why am I writing this book?	1-1
1.2	An essential tension	1-2
1.3	A few words on names	1-3
2	**Cecilia Payne**	**2-1**
2.1	Beyond Curie	2-1
2.2	No insuperable objections	2-5
2.3	The Harvard Computers	2-7
2.4	Starstuff	2-15
2.5	Two astronomers from Cambridge	2-25
2.6	Reactions	2-27
2.7	Blocked paths	2-28
2.8	Love (of science) levels all ranks	2-30
2.9	Science summary: stellar spectra	2-38
	References	2-42
3	**Lise Meitner**	**3-1**
3.1	Making up for lost time	3-1
3.2	Questions of credit	3-4
3.3	A scientific powerhouse	3-6
3.4	Tumult	3-7
3.5	How Nobel Prizes are selected	3-13
3.6	Beyond uranium	3-15
3.7	The breakdown of science	3-20
3.8	Our Madame Curie	3-22
3.9	Science summary: nuclear fission	3-23
	References	3-27
4	**Chien-Shiung Wu**	**4-1**
4.1	Mighty hero	4-1
4.2	Exile	4-3

4.3	Pushing back	4-6
4.4	Rising through the ranks	4-9
4.5	'Wasting her time'	4-10
4.6	Instant Nobel	4-18
4.7	Honors	4-25
4.8	Science summary: parity	4-25
	References	4-30

5 Maria Mayer   5-1

5.1	The seventh generation	5-1
5.2	From nuisance to necessary	5-3
5.3	A new era	5-9
5.4	Magic	5-11
5.5	A different way to win a race	5-13
5.6	Quarter loafs	5-16
5.7	A typical genius	5-17
5.8	Science summary: nuclear shell model	5-18
	References	5-23

6 Afterword   6-1

Acknowledgments

First and foremost, I want to acknowledge my fiancée Erin Eisenbarth, who came up with the idea for the book and has watched me wrestle with it all the way along. She is a historian, and I am a physicist, and I found her insights on how to do historical research invaluable. On occasion, I asked her to comment on sections or help me understand some thorny historical problem; at other times, she would encourage me not to allow myself to be too distracted with minutiae far removed from the main narrative. While I don't have the same kind of expertise to offer her as she completes her own dissertation, I hope I can at least help her in return by cooking us dinner, and perhaps baking the occasional batch of brownies, as she reaches crunch time in her own project.

I'd also like to acknowledge several former students of mine, each of whom is now a scholar in her own right: Jing Min Chia, who performed much of the investigation of Mayer's time at Sarah Lawrence, working with Abby Lester, the superb archivist at Sarah Lawrence College; Kristen Koopman, who invited me to Virginia Tech to give a seminar based on my preliminary work on the chapter on Cecilia Payne; and the historian of science Dr Kelly O'Donnell, who provided me valuable feedback on that same chapter.

One thing I have discovered about this process is that writing about history is hard. The historical record is full of apparent contradictions; modern writings often get taken in by legends that have grown up over time, or misinterpret key pieces of evidence. While I have uncovered a few instances where that has occurred, I am sure I have fallen prey to other misconceptions, and likely created a few new misinterpretations of my own. I look forward to hearing from readers who know better about one point or another, and themselves want to set the record straight.

Author biography

Scott Calvin

With a major in astronomy and physics and a minor in classics from the University of California Berkeley, followed by a PhD in chemical physics from the City University of New York, Dr Calvin has taught, conducted research, and advised students in a wide variety of institutions, including the Naval Research Laboratory, Brookhaven National Laboratory, the Hayden Planetarium, the Stanford Synchrotron Radiation Laboratory, Sarah Lawrence College, Examkrackers, and Lehman College. His course offerings have run the gamut, including innovative courses such as Crazy Ideas in Physics, Rocket Science, and Steampunk Physics, as have his books, including *We Can Do It! A Problem Solving Graphic Novel Guide for General Physics, XAFS for Everyone*, and the open-source artisanal pop-up book *National Synchrotron Light Source II: Long Island's State of the Art X-Ray Microscope*.

IOP Concise Physics

Beyond Curie
Four women in physics and their remarkable discoveries, 1903 to 1963
Scott Calvin

Chapter 1

Introduction

1.1 Why am I writing this book?

People who know me are not generally surprised that I am writing this book; I have a wide range of interests, including the history of science and extending science education to underrepresented groups. But for some of those who just met me, there is a bit of surprise or even skepticism: a cursory scan of the author list for books about women in science shows that it's mostly (but not exclusively!) populated by women. Since I am not a woman, it could seem that there must be some special motivation that is driving me to write this book.

The truth, though, is there is no single reason or inspiring incident that led me to this topic. Instead, I can think of several inspirations that led me to write the book you now hold in your hands or see on your screen.

Ever since I read Ruth Lewin Sime's biography of **Lise Meitner**, Meitner has been a scientific hero of mine, a role model for experimentalists dedicated to the practice of science.

Soon after joining the faculty at Sarah Lawrence, I became aware that **Maria Mayer** had once taught there, and had gone on to win a Nobel Prize. At the time, I found that mildly interesting, but not compelling. But years before, as an undergraduate, I had greatly enjoyed the science fiction novel *Timescape*, written by the physicist Gregory Benford, and so I assigned it to my students in my freshman seminar. Upon re-reading it, I was delighted to find that a fictionalized Mayer appears in it as a character. Benford attended UCSD while Mayer was there, and had set a portion of the book during that time. Since many of the characters were fiction, I hadn't even been sure Mayer was a real person the first time I read the book, and I certainly hadn't remembered her name—but I did remember her key moment, and how it struck me. In *Timescape*, Gordon Bernstein, the main character, also a physicist, had come to a surprising conclusion about an experiment he was conducting, with the result that he was getting a lot of pushback from other physicists. The fictional Mayer asks Bernstein a question: 'Do you believe your results?' When Bernstein replied that he did, Mayer replies with a single word: 'Good.' That small affirmation is a lifeline for Bernstein, and I found it very moving.

One of the students in my freshman seminar, Jing Min Chia, also took an interest in Mayer. At first, she was researching women scientists in general, but Mayer, with her Sarah Lawrence connection, was a good place to start. Working with Abby Lester, Sarah Lawrence's masterful archivist, she soon found that many of the capsule biographies of Mayer were misleading, implying that Sarah Lawrence had considered her a 'nuisance'. Chia decided to refocus her investigation entirely on Mayer, and ended up writing an eight thousand word biography covering Mayer's time at Sarah Lawrence to set the record straight.

Chia's work was remarkable, particularly for a freshman, and now has the distinction of having been cited in the book Mayer's son Peter wrote about her life, on the website of the Sarah Lawrence archives, and in this book. In 2012, Sarah Lawrence held an early commemoration of the 50th anniversary of Mayer's Nobel, with a panel discussion which included, among others, Peter Mayer and Chia. Ironically, Mayer, who lived in Guam, happened to be in the United States at the time and attended in person, while Chia, who was a student at Sarah Lawrence, Skyped in from Malaysia.

Despite the attention it has received, Chia's work has never been published in full, a status I hope she will correct some day.

While I had read the autobiography of **Cecilia Payne** some years before, her story did not really strike me until I began to teach a year-long astronomy seminar at Sarah Lawrence. It was then that I became entranced by the Harvard Computers, the women who were responsible for so much of the progress of astronomy in the late 19th and early 20th centuries. I could see how Payne represented the next step of progress for women in the field, and how the torch was handed from one generation to the next.

And so when Jeanine Burke of IOP reached out to me to see if I had any ideas for a book in the Concise Physics series, I shared the query with my fiancée, Erin Eisenbarth, a feminist historian, who instantly thought of my interest in these women, and thought I should write a book about them.

And so I wrote a book about them.

1.2 An essential tension

All of the physicists in this book are scientists. In most contexts, I would no more refer to Meitner as a 'woman scientist' than I would to Newton as a 'man scientist.' (For that matter, despite the great anti-semitism Einstein faced, I do not usually use the term 'Jewish scientist' to describe him.) To do so, it seems to me, risks implying that there is such a thing as 'women's science,' a description the four scientists in this book would certainly dispute; Chien-Shiung Wu, for example, did so explicitly and repeatedly.

Lise Meitner is not a hero of mine *because* she is a woman, nor did I admire the Maria Mayer I first met in fiction for that reason. But how society and the scientific community treated them was certainly influenced by their gender, and their responses to the challenges they faced add to my admiration for them.

But now I've placed them together in a book, so that they are tied together both by profession and gender. If I don't want to write about 'women physicists,' how am I to write about these four physicists who were women?

I decided, in each case, to focus on the event in each one's scientific career that she is best known for. Sometimes that happened at the start of a long, successful career, as was the case with Payne; sometimes near the end of one, as happened with Meitner. Using an event as a focus provides structure: how did each of these scientists get to that point? What obstacles did she face, what decisions did she need to make? Who were here allies, her rivals, and her detractors? How was the work recognized at the time? These questions could be asked of any scientist making a major discovery. Some aspects of the answers have little to do with the gender of the scientists in question, while others are due entirely to the fact that each was a woman.

Because of this structure, I didn't feel compelled to discuss, for example, the love life of each, or their deaths, or their physical appearance. But when one of those aspects was germane to the point at hand, I discussed it as a part of the context.

Some issues calling for extended discussion were likely to arise repeatedly: the process for awarding Nobel Prizes, for instance, or the question of child-care. In these cases I provided the primary discussion within one of the four chapters only. This has the side-effect that the chapters don't stand as well separately as they do together.

I also wanted this book to be helpful for people coming from a wide variety of backgrounds: among them scientists who want to know some history and historians who want to know some science. In order to serve those various audiences, and to provide some differentiation between this book and others that touch on the same historical figures, I've provided a science summary at the end of each chapter. For those who aren't interested in the science itself, these can be safely skipped. On the other hand, for those who want to understand the science before trying to unpack the history, they could be read first.

1.3 A few words on names

Three of the four physicists featured in this book at least experimented with taking their husbands' names upon marriage, raising the question as to how to refer to them in this book.

My guiding principle was to use the name under which they published the key work or works discussed in this book. This leads to modestly unconventional choices in two cases.

I write 'Cecilia Payne' (usually shortened to 'Payne'), because that was the name she used for her famous dissertation. Most modern authors, however, choose to use Cecilia Payne-Gaposchkin, the name she used after her marriage, both personally and professionally.

Maria Mayer is the trickiest case. Born Maria Gertrud Käte Göppert, she shared a first and last name with her mother, although her middle names were different. After marriage, she published some papers as Maria Göppert-Mayer, with the modestly Americanized spelling Maria Goeppert-Mayer also appearing. As her career progressed, though, she dispensed with the hyphen, leaving her full name as Maria Goeppert Mayer, sometimes written Maria G Mayer or simply Maria Mayer. Many modern accounts use Maria Goeppert-Mayer, including portions of the official Nobel Prize website. While not entirely wrong, as she did use that name a few times early in her career, it is not the way she referred to herself or published either at the time she did the work which won her the prize or at the time she was awarded it.

In addition, Maria's husband Joe Mayer plays a more important role in her scientific career then did the husbands of Payne or Wu. This would make referring to her as 'Mayer' throughout the text potentially confusing; perhaps this is why there is a preference for Goeppert-Mayer in many modern writings. My solution, admittedly imperfect, is to refer to Joe and Maria by their first names through most of their chapter, and to use 'Mayer' to refer to Maria Mayer in other chapters and in the science summary, where there is no danger of confusion with Joe. I have made this choice to honor her eventual preference for her own last name ('Mayer'), while still maintaining clarity in the text, and not treating her differently because of her gender (when she is 'Maria,' her husband is 'Joe').

The remaining cases are straightforward. Chien-Shiung Wu used Wu in her professional life, and Mrs Yuan (her married name) socially, being careful to maintain a rigid distinction between the two. Accordingly, she is 'Wu' in this book.

Lise Meitner maintained the same name throughout her life.

IOP Concise Physics

Beyond Curie
Four women in physics and their remarkable discoveries, 1903 to 1963
Scott Calvin

Chapter 2

Cecilia Payne

2.1 Beyond Curie

In 1903, or thereabouts, a little English girl was being taken for an evening stroll in her baby carriage. A shooting star flashed across the sky; her mother told her it was a meteorite. It was the little girl's first taste of astronomy [1].

That girl, Cecilia Payne, had been born in the same year as the first Nobel Prize. Around the time she was shown the meteorite, the 1903 Nobel Prize in Physics was awarded to Henri Becquerel, Pierre Curie, and Marie Curie. Since the Nobel Prize itself was only three years old, it is a stretch to consider this a shattering of barriers. And in fact, it did not herald a new era of equality. When Marie Curie won her first Nobel, it meant that 5% of prize-winners, in all fields, had been female. As of this writing, in 2016, the fraction of female Nobel prize-winners has risen…to 5.4% [2].

Eight years later, in 1911, Marie Curie won again, this time being the sole recipient of the Nobel Prize for Chemistry. It was the first time anyone had won the Prize twice.

Cecilia Payne, the little English girl, was no longer quite so little. At school she had learned French, German, and Latin. She enjoyed solving quadratic equations and knew how to use a chemical balance.

It was not terribly unusual for a girl to develop those talents. Marie Curie, while certainly a topic of conversation, was not the only female scientist the young girl was likely to have heard of. The previous century had made celebrities of several: notably, the astronomers Maria Mitchell in the United States and Caroline Herschel in Germany, as well as the polymath Mary Somerville in Scotland.

Somerville, in particular, filled the role in the popular imagination that Curie would later occupy, particularly in the United Kingdom. An expert in mathematical astronomy and astrophysics, she also delved into geology, geophysics, ecology, chemistry, and biology. In 1834 she wrote a book *On the Connexion of the Physical Sciences* [3]. 'Astronomy', she wrote in its introduction, 'affords the most extensive example of the connection of the physical sciences. In it are combined the sciences of number and quantity, of rest and motion'.

In an early review [4] of Somerville's book, the philosopher of science William Whewell noted that Somerville's emphasis on the relation between the sciences was badly needed, as there was at that time no English word for 'students of the knowledge of the material world collectively'. Whewell refers to discussions in the British Association for the Advancement of Science, where labels were suggested and then rejected, including 'philosopher' (too broad), 'savans' (too French), and 'nature-poker' (too silly). 'Scientist' was also suggested, but at the time was not considered 'palatable'.

Whewell's work is the first time that the word 'scientist' is known to occur in print. It gradually gained currency, and would eventually be used to describe Somerville herself. While the term was not invented to describe her, nor was it invented by her, it was her work that gave an occasion for its introduction. It is reasonable, then, to describe Somerville as one of the inaugural class of scientists; the first group of scholars to be graced by that term during their lives.

On the Connexion of Sciences was a best-seller [5], and Somerville rapidly became famous. A sailing ship was named after her, with her likeness for a figurehead [6]. A famous sculptor was commissioned to create a bust of her to reside in the Royal Society [7] (figure 2.1)—an irony, since her sex debarred her from membership of that scientific organization—and the explorer Admiral Parry named an island after her. Shortly after her death, Oxford opened a college for women named Somerville Hall.

Mary Somerville was once a household name, but that is no longer the case, at least in the United States. Partially, that may be an effect of the passage of the time, and partially it may be that there is no single discovery or insight that encapsulates Somerville's career—no apple falling from the tree, discovery of radium, or realization of the double-helix structure of DNA. Another likely factor, however, was the rise of Marie Curie, who did not desire fame ('In science we must be interested in things, not in persons' [8]), but found herself drowning in it nonetheless. In 1912, following her second Nobel, the American papers anointed Curie 'without doubt the greatest of all women scientists' [9].

This praise of Curie marks a milestone of sorts. Somerville's scientific accomplishments placed her, in the words of one speaker, 'in the foremost ranks of modern physicists and speakers' [10]. While there was frequently discussion, and all manner of opinions, on the relationship between her gender and her scientific ability, it would have seemed strange to create a category ('women scientists') in which to rank her (figure 2.2).

By Curie's time, however, the profession of scientist was well-established, and there were just enough women in it to allow comparison within the group. Then, once a member of the public was told who is the greatest in what is implicitly a lesser category, why bother with the rest? The words used to praise Curie inadvertently swept aside other scientists who shared her gender.

This can be seen in figure 2.3, which shows the relative frequency of appearance of Marie Curie, Mary Somerville, and Caroline Herschel in American writings over

Figure 2.1. Bust of Mary Somerville at the Royal Society. © The Royal Society.

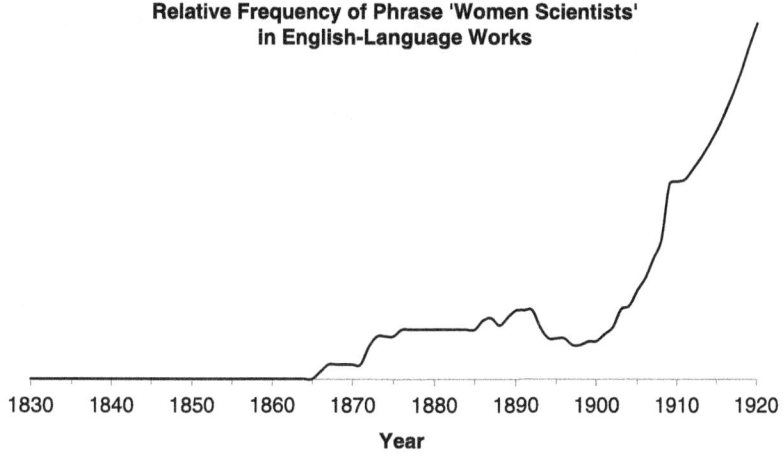

Figure 2.2. Google ngram frequency for phrase 'women scientists,' using smoothing of ten. 'Female scientists,' 'lady scientists,' and 'woman scientists' all had a negligible frequency during this period.

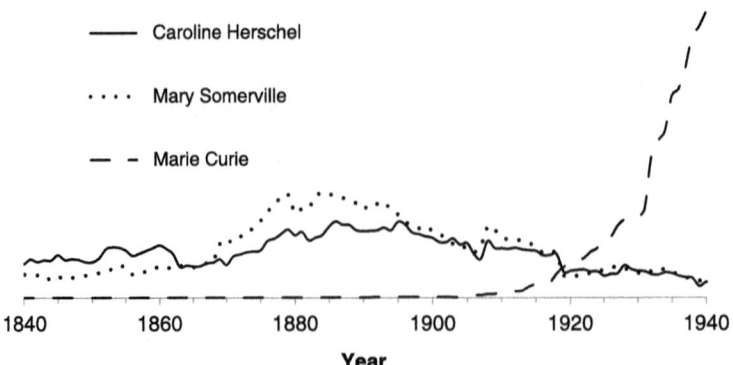

Figure 2.3. Google ngram frequency for each scientist's name, using smoothing of five. Note the sharp drop for Somerville and Herschel around 1920, when Curie was becoming a superstar.

time. Curie's gains in the early 20th century correspond to a sharp drop in references to the earlier scientists[1].

Payne, 11 at the time of Curie's second prize, would have known that science was not an impossible dream. She had already taken a private vow (to a spruce tree and an orchid!), years before, to dedicate herself to the study of nature. She set about making that dream not just a reality, but a passion, sneaking in to the small science lab at school:

> The chemicals were ranged in bottles round the walls. I used to steal up there by myself (indeed I still do it in dreams) and sit conducting a little worship service of my own, adoring the chemical elements. Here were the warp and woof of the world, a world that was later to expand into a Universe. As yet I had caught but few glimpses of it—the meteorite, Halley's Comet, the Daylight Comet of 1910. I had yet to realize that the heavenly bodies were within my reach. But the chemical elements were the stuff of the world. Nature was as great and impressive to me as it had seemed when I stood under the spruce tree and vowed myself to its service [1].

A few more years would pass, and Payne would become a young woman, freshly accepted in to Newnham College at Cambridge. She was ready to follow her dream.

---

[1] Maria Mitchell, an American, does not suffer the same fate in American writings. Likewise, the Scottish Mary Somerville persists into the Curie era much more strongly in British works. There is apparently space for national heroes alongside Curie.

2.2 No insuperable objections

When Payne entered Cambridge as a student of its all-female Newnham College in the fall of 1919, it was a time both of great optimism and, particularly for women, of great frustration. In her autobiography, Payne chose to emphasize the former, while not entirely neglecting the latter:

> The atmosphere was euphoric. The 'war to end war' was over…we women, of course had no votes (even had we been old enough), but that did not prevent us from conducting spirited debates. A new world was opening before mankind [1].

Ernest Rutherford had recently been brought back to Cambridge to direct the Cavendish Laboratory, and Cambridge Observatory Director Arthur Eddington, recently returned from his eclipse expedition testing Einstein's general theory of relativity, was on the verge of becoming a superstar in the popular press [11]. While women had been attending Cambridge for nearly half a century, they were not yet permitted to earn degrees or participate in its governance, despite repeated attempts to remove these prohibitions. The issue once again came to a head during Payne's time there, with a vote on both questions taking place in 1921. The votes established a compromise position: women could earn 'titular' bachelor's degrees, but without the governance rights awarded to men. A contemporary editorial [12], while purportedly on the side of granting governance rights to a limited number of women, was also sympathetic to the arguments of the opponents: to wit, that granting those rights meant 'Cambridge would be destroyed as a first-rate University for men', and that what women wanted was 'not to be fairly treated, but to get power in the University'. The editorial went on to provide sympathy to those who advanced those arguments, saying that male students found it 'just now almost as easy to hate the woman whom he imagines to be an unscrupulous rival as, in another mood, he might find, [sic] it easy to flirt with her'. Many of these male students, buoyed by their victory in the Cambridge Senate, celebrated by rioting in front of Newnham, badly damaging its historical gates, which were a memorial to Newnham's first principal, the suffragist Anne Jemima Clough (figure 2.4). Payne, living behind those damaged gates and passing through them on her way to lectures in other parts of Cambridge, would have faced a daily reminder of the vitriol of those who would deny her the rights they held. Not that she would need the gates to remind her. As the editorial also states, the sentiment at Cambridge for keeping women out, or at least for keeping them in a second-class status, 'was largely guided by the medical and scientific men, who resent the way in which their places are usurped by women in laboratories'. Women, that is, like Cecilia Payne.

Students of natural science at Cambridge had to choose three subjects for their initial focus; Payne chose the unusual combination of botany, physics, and chemistry. At first, she believed she would specialize in botany when the time came to choose, but her interest and confidence in that subject were waning. Fortuitously, near the end of her first term she scored a ticket to Eddington's lecture on relativity and the results of his eclipse expedition. Payne was entranced; by her

Figure 2.4. Damaged gates at Newnham College. With permission from Newnham College, Cambridge.

recollection she was so shaken by the ideas that she did not sleep for three days. From that point forward, Payne committed herself to astronomy.

Not that she could simply do so. She had begun a course of study in the natural sciences. For historical reasons, astronomy at Cambridge was considered a branch of mathematics instead. So Payne chose to direct herself toward the study of physics, while attending as many lectures on astronomy as she could.

Before long, she reached the point where she was attending the advanced physics lectures given by the esteemed Rutherford. Regulations in place at the time required that women be segregated in the front row. Since she was the only woman attending, this made her very conspicuous. To make matters worse, Rutherford ritually humiliated her (regardless of what his intention may have been, that was the effect) at the start of each lecture, to the great amusement of the rest of the students.

That was physics, however, and what Payne most wanted was astronomy. One night when the Cambridge Observatory was having an event for the public, she came and so peppered the assistant manning the telescope with questions that he fled the scene. The assistant found Eddington and brought him back to help with Payne's questions, either by answering them or finding some other way to deal with her. By the time Eddington got there, Payne had taken over the demonstration, and was giving an extemporaneous lecture to the assembled visitors as she helped a young girl to look through the telescope. Taking in the scene, Eddington chuckled, and it was only then that Payne realized her idol had come in. It was now or never.

'I want to be an astronomer'.

Payne's recollection of the exact sequence of events that followed was a bit hazy, despite—or perhaps because of—the importance she later placed on them. That leaves us free to imagine the moment, somewhat as Payne later did: Eddington considered the young woman in front of him who had, for a few minutes, taken over the duties of his assistant after asking him questions for which he had no answer, and was now giving her own answers to the public, using the Observatory's equipment. A brief pause, a raised eyebrow, and then Eddington's reply:

'I can see no *insuperable* objection'.

Eddington gave her the titles of books that would be helpful to read, only to find that Payne had read them all. Next, he gave her access to the Observatory library. Attendance at his lectures and his teas followed, and then research under his direction and her first publication, a four page paper on the proper motions of stars in and near the line of sight of the open cluster M36 [13].

But in those days, there was no way forward for a middle-class woman to become a professional astronomer in England[2]. The only career she saw open to her was that of schoolteacher. She needed a way out.

Her chance arrived in the person of Harlow Shapley, the new Director of the Harvard College Observatory (HCO), who came to London to deliver a lecture in 1922. Securing an introduction from a mutual acquaintance, Payne asked if she could come to work under him at the HCO.

Shapley's reply was flippant and facile—of course she could come…and then take over from Annie Jump Cannon, once Cannon retired. Cannon was then the leading member of the female astronomical staff known as the Harvard Computers, with her own title (Curator of Astronomical Photographs), numerous publications, and an international reputation, including an honorary doctorate recently awarded by the University of Groningen. Suggesting to Payne that she might be Cannon's eventual replacement would be like suggesting to a cub reporter for a newspaper that he might be the next editor—not impossible, but not likely, and it would involve jumping over a lot of other people.

Payne, however, heard the answer as 'yes', and proceeded to make it happen. She secured enough fellowships and grants to cover a year in the United States. Unlike many similarly-trained women of the time, she wasn't just looking for a job or for a way to satisfy a passion for astronomy. Cecilia Payne was looking to become a professional astronomer.

2.3 The Harvard Computers

Before we follow Payne's journey across the Atlantic, it is important to discuss the history of the community she would find when she arrived: the Harvard Computers.

As recently as 1971, the Oxford English Dictionary still defined a computer without any reference to the modern meaning of the word, solely as 'one who computes; a calculator, reckoner; *spec.* a person employed to make calculations in an observatory, in surveying, etc' [14]. The last sense given in the definition indicates

---

[2] That's not to say that there weren't female professional astronomers *in* England. For example, while Payne was still at Cambridge, A Vibert Douglas came from Canada to work with Eddington, in much the same way that Payne would go to America to work with Shapley. But Douglas soon returned to McGill University, in Montreal, to secure her doctorate [15], while Payne would remain in Massachusetts for the rest of her career.

it is a job title, or profession. In our current vernacular, there are many occupations where practitioners, either routinely or in exceptional cases, go beyond the literal meaning of their job title. For example, we are not surprised to hear that a 'baker' has devised a new recipe, and we certainly expect bakers to spend part of their time mixing, and kneading, and beating eggs, even though none of those things involve the literal meaning of 'bake'. Similarly, editors may solicit and evaluate contributions in addition to editing them, and sailors do more than raise and lower sheets of cloth on oceangoing vessels. And yet, in each case, defining the name of a profession by a relatively narrow task hints at a lower prestige. Identifying a profession by a field of operation, in contrast, lends an air of importance: 'lawyer', 'politician', 'artist', 'scientist'.

Thus, in the sciences in the early 20th century, the people called computers often did much more than rote calculation, but, like the baker whose recipe is adopted by a famous chef, were thought of as being on a lower level than the scientists they worked for.

The Harvard College Observatory (HCO), like most major observatories, employed a staff of computers to help process and analyze the voluminous data collected by the Harvard telescopes, particularly now that observations could be captured on photographic plates for later characterization. But unlike most observatories, the computers were all women, and had been for decades.

How did this come to be? What kind of work were these women allowed to do, and what recognition did they get both within the Observatory, in the scientific community at large, and in public discourse? If we are to understand Payne and her work, we should first make an attempt at addressing these questions.

The story, as it is often told now, is more or less this: in 1881, Edward C Pickering, the new Director of the HCO, was dismayed by the poor work being done by his male computers. In frustration, he swore that his Scottish maid could do a better job...and for half the pay. What had begun in a moment of anger became a plan: he fired his team of male computers and brought in a team of women, saving the Observatory a good deal of money and establishing a tradition that would last for decades. Informally and rudely, others often referred to the women as 'Pickering's Harem'.

But while some of the individual components present in the previous paragraph are based on evidence in historical documents, as a coherent narrative it does not stand up to scrutiny. For one thing, there were female computers at the Observatory *prior* to Pickering's arrival. And while several members of the staff were in fact let go, this did not occur for several years after Williamina 'Mina' Fleming, the 'Scottish maid', was hired [16].

Aside from the chronological difficulties, core elements of the story did not appear in writing for half a century. The earliest use of 'Pickering's Harem' that I could find reference to was in 1976 [17]. It would be unsurprising to find that some contemporary of Pickering's, in an attempt to be witty or derogatory, had once used the term. But if, as is often depicted now, it had been a routine smear, I would expect to find reference to it in the recollections of those who served at the Observatory at the time or soon after—many were not shy about disclosing other slights!

Likewise, the notion that Pickering decided to hire Fleming because of frustration with a male computer first appears in print, as far as I can find, in a short biography of Fleming written by Dorrit Hoffleit (*Notable American Women 1607–1952*), herself a prominent astronomer who came to Harvard not long after Payne. Hoffleit, who never met Pickering or Fleming and came to the Observatory more than four decades after the event in question, accompanies it with the caveat 'so the story goes' rather than attributing it, as is her habit with other information, to a particular source. This suggests it may have been a legend that gradually gained currency.

Finally, there is the question of pay. The starting salary for Pickering's female computers is known: 25 cents per hour, seven hours a day, six days a week, eleven months a year. While it is somewhat problematic to try to convert dollar amounts from the late 19th century into modern values (manufactured goods, for example, would have been relatively more expensive then, and real estate relatively less) one calculation would suggest that 25 cents per hour in 1881 is roughly equivalent to $6 per hour now. Many of the computers did not have a college education, but even so, that suggests a value below what we would now consider a living wage[3]. On the other hand, that was the going rate for work of that type at that time. As a contemporary source noted [18], the women were:

> employed not from the meaner motive which so often leads to the opening of some new field for women's work, viz., that their work can be obtained at a cheaper rate than that for men; for the women [sic] assistants doing routine work are paid at the same fixed rate per hour as the men in other departments of the Observatory who do the same kind of work.

But while technically accurate, this statement is somewhat disingenuous. Although it is true that entry-level men and women were paid the same, the women received very little in the way of raises even as their responsibilities increased. In fact, the same source says that the work done by women at the Observatory could be divided into three classes: computing, 'original deductions', and the exhaustive cataloguing and classifying of stars for which the Observatory soon became famous.

In the early years, Pickering saw the use of a large team of computers as a matter of efficiency, but not because of their gender. He felt that the time and effort of experienced astronomers were being wasted on tasks that could be done by those with much less education and skill. In his annual report to the President of Harvard [19], he put it this way:

> In large observatories it is not unusual to establish a number of departments, each under the entire charge of an astronomer who is often unaided by assistants…there is often a lack of cooperation,—it is a ship in which all the sailors are captains. It is not clear that better results are thus obtained with a

---

[3] While some working women in the 19th century were married, and thus could rely in part on their husbands' incomes, that was not generally the case for the Harvard computers, most of whom were single, often remaining so throughout their life. Low salaries, therefore, represented a significant hardship.

given expenditure of money, than if assistance was given to amateurs who had displayed especial skills in their work...[But at the HCO] many of the assistants are skilled only in their own particular work, but are nevertheless capable of doing as much and as good routine work as astronomers who would receive much larger salaries. Three or four times as many assistants can thus be employed, and the work done correspondingly increased for a given expenditure.

The last sentence above is sometimes taken out of context by modern writers to suggest Pickering was hiring women instead of men so that he could pay them a fraction of the amount. In fact, he was hiring mid-level computers and assistants rather than broadly-trained astronomers, arguing that this yielded greater efficiency.

As early as 1879 this model was spoofed by a member or friend of the Observatory staff, who rewrote Gilbert and Sullivan's *HMS Pinafore* as *The Harvard University Pinafore*[4]. In Gilbert and Sullivan's opera, Josephine, a Captain's daughter, is torn between her love for a common sailor and her duty to marry an admiral. In the end, it is revealed that the sailor was switched with the Captain as a baby, and is thus actually of a higher lineage than Josephine, allowing them to marry, while simultaneously making her marriage to the Admiral socially unacceptable. Love levels all ranks 'to a considerable extent', says the Admiral, 'but it does not level them as much as that' [20].

In the HCO version, Josephine conveniently becomes Joseph, as there was a young assistant at the HCO of that name at the time. Rather than having to choose between suitors for marriage, he, as a promising young talent, must choose between two possible employers. 'Love levels all ranks' remains in the script, but it now refers to a mutual love of astronomical work, allowing those from different background to work together [21]. The Computers, by this point already composed of women, represent the members of the crew, and Pickering its captain—it is possible that this is the source of Pickering's imagery that other observatories are like 'a ship in which all the sailors are captains', although the analogy is natural enough even if Pickering remained unaware of the script.

But why did Pickering choose to form a team made up exclusively of women, instead of using semi-skilled men? One reason was connected to the low pay and status. While he wasn't paying women less than men for positions at the same level (at least when they were first hired), it was easier to find women, who had fewer opportunities elsewhere, willing to work for those sums. In fact, the Superintendent of the United States Naval Observatory, defending himself and his agency against charges of overspending, called Harvard out on this point [22]:

---

[4] The authorship of the piece is uncertain. When the manuscript was rediscovered at the HCO in 1921, a bit of amateur detective work suggested that the author was Winslow Upton, a member of the staff at the time of writing. But further investigation in the late 20s revealed the handwriting was not his, and may in fact have been Mina Fleming's. The working explanation then became that it had been written by Upton and then copied by Fleming, but the evidence for Upton's involvement is flimsy, particularly since his widow didn't know of it [23]. This raises the intriguing possibility that Fleming, or perhaps one of the other computers, was the actual author.

It is in the lower grades particularly, however, that I ask for a fair comparison of salaries…at Greenwich [Royal Observatory], for example, computers receive an average of $325 per year, less than one-half the pay of our laborers, and less than the renumeration of any human being doing skilled work in the United States. At Harvard computers receive an average of $600, less than the pay of any person in the United States Service. These computers are largely women, who can be got to work for next to nothing.

Actually, at Greenwich, the experiment had been tried of hiring college-educated women as computers, but it was found that the salaries were insufficient to recruit and retain women, and the plan was soon abandoned. At the time the Superintendent was writing, that role was filled instead primarily by high school boys [24].

But there was also another, more progressive, argument that was in the air at the time, to which Pickering occasionally made allusions. And that argument came from another astronomer—the most famous female astronomer in American at the time, Professor Maria Mitchell. In 1876, five years before Pickering added Mina Fleming to the staff of the Observatory, Mitchell gave an address at the Fourth Congress of the Association for the Advancement of Women entitled *The Need of Women in Science* [25]:

When I see a woman put an exquisitely fine needle at exactly the same distance from the last stitch at which the last stitch was from its predecessor, I think what a capacity she has for astronomical observations. Unknowingly, she is using a micrometer; unconsciously, she is graduating circles. And the eye which has been trained in the matching of worsteds is specially fitted for the use of prism and spectroscope. Persons who are in charge of the scientific departments of colleges are always mourning over the scarcity of trained assistants. The directors of observatories and museums not infrequently do an immense amount of routine work which they would gladly relinquish. Their time and strength are wasted on labor which students could do equally well, if students could be found who would be ready to make science a life work.

Mitchell counted the first and second Directors of the HCO among her friends [26]; she worked for the third for a time as a computer when they were both with the United States Coast Survey [27]; Pickering was the fourth. He would have been well aware of her opinions on the matter.

Indeed, Pickering's own language echoed Mitchell's:

Much valuable assistance might be rendered by a class whose aid in such work has usually been overlooked. Many ladies are interested in astronomy and own telescopes, but with two or three noteworthy exceptions their contributions to the science have been almost nothing. Many of them have the time and inclination for such work, and especially among the graduates of women's colleges are many who have had abundant training to make excellent observers. As the work may be done at home, even from an open window, provided the room has the temperature of the outer air, there seems to be no reason why they should not thus make an advantageous use of their skill. It is believed that it is only necessary to point the way to secure most valuable assistance [28].

In any case, in the early days of his Directorship Pickering clearly envisioned the women as semi-skilled labor, freeing the more-educated male astronomers from routine work so that they could spend their time more productively. In what is, to modern ears, one of his more uncomfortable essays, he compares the skill level needed by computers, and thus the appropriate pay, to that of beaters in tiger-hunts in India. Specifically, he outlines their duties as 'copying numbers on prearranged forms, and computing in which only a knowledge of the four rules of arithmetic is needed' [29].

Soon, however, their duties began to progress beyond that, particularly in the case of Mina Fleming, the 'Scottish maid', who increasingly carried out sophisticated investigations and analyses. At first, Pickering stuck to his factory model, announcing Fleming's results in the *Annals of the Harvard College Observatory* without specifying authorship. Beginning in 1886, Fleming led the work on the ambitious Draper Catalogue, including the classification of more than ten thousand stars using a system that Fleming herself invented. When the results were published in the *Annals* in 1890, the title page lists only Pickering, but on the second page of the introduction he writes that 'nearly all the measures described in this Volume were made by Mrs M Fleming, who also superintended their reduction, and rendered important aid in preparing the work for publication' [30]. This makes it clear that, by any reasonable standard, Fleming was an author of the work, and most likely the lead author, despite not being given that credit.

But Fleming's role did not escape the publishers of the *Observatory,* a journal associated with the Royal Observatory at Greenwich, England. In a review of the Draper Catalogue, they wrote [31]:

It would be difficult to say too much in praise of the zeal and skill with which this great work has been accomplished. The name of Mrs Fleming is already well known to the world as that of a brilliant discoverer; but the present volume shows that she can do real hard work as well.

An extended feature on the Harvard Computers in *New England Magazine* followed [18], asserting that 'this corps of women, in addition to doing thorough routine work, has shown great capacity for original investigations'. While this publication was careful not to explicitly refer to the computers and assistants as 'astronomers', the avalanche of coverage that followed in the American and international press made no such distinction. Mina Fleming, to the popular mind, was now no longer just a computer, or even an assistant. Mina Fleming had become an astronomer.

Pickering gradually adapted to the reality that, rather than his team consisting solely of women computers with low, but arguably fair, pay, it now included at least one desperately underpaid astronomer of international stature. The 1897 follow-up to the Henry Draper Catalogue indicates on the title page that it is 'by' Pickering, who was 'aided by' Fleming [32].

When Pickering hired Fleming, she was a diamond in the rough, a high-school educated school-teacher who had fallen on hard times [33]. As Fleming's abilities grew, Pickering began to hire women with college educations in the sciences, among them Antonia Maury (Physics, Astronomy, and Philosophy at Vassar under Maria Mitchell) [34], Henrietta Leavitt (Radcliffe), and Annie Jump Cannon (Physics at Wellesley) [35], each of whom became famous astronomers in their own right.

By 1898, the question of credit reached a tipping point. A conference of astronomers and astrophysicists was held at Harvard, with the additional goal of creating the society that would become today's American Astronomical Society [36]. 28 papers were given, including two by Fleming and one by Maury, each under their own name and without Pickering as coauthor. According to a contemporaneous account in *Popular Astronomy* [37], the first of Fleming's papers was read to the assembled astronomers, by Pickering, perhaps because of Fleming's unusual status as a paper author without a college education (Maury presented her own paper, so gender was not the sole reason Pickering presented Fleming's work). Pickering could have left it at that, and my intuition is that the Pickering of a decade earlier would have. Instead, he felt compelled to add that Fleming's paper didn't mention that she herself had been the discoverer of almost all the stars discussed in the paper, a startling feat obscured by the passive construction favored in HCO scientific publications up until that point. The room burst into applause, prompting Mina Fleming to arise from the audience and take her place at the front of the room to answer questions about the work.

In 1899, Fleming became the first woman to be given a formally recognized position at Harvard (an 'officer of the College'), as 'Curator of Astronomical Photographs'. Her salary was now greater than that of an entry-level computer, but lagged behind that of men with comparable responsibilities. The next year, Fleming wrote this in a journal intended for a time-capsule [38]:

> I had some conversation with the Director regarding women's salaries. He seems to think that no work is too much or too hard for me, no matter what the responsibility or how long the hours. But let me raise the question of salary and I am immediately told that I receive an excellent salary as women's salaries stand. If he would only take some step to find out how much he is mistaken in regard to this he would learn a few facts that would open his eyes and set him thinking. Sometimes I feel tempted to give up and let him try someone else, or some of the men to do my work, in order to have him find out what he is getting for $1500 a year from me, compared with $2500 from some of the other [male] assistants. Does he even think that I have a home to keep and a family to take care of as well as the men? [Fleming was a single mother.] But I suppose a woman has no claim to such comforts. And this is considered an enlightened age!

While her salary still languished, her reputation soared, aided by Pickering's promotion of her among the astronomical community [16]. In 1905, eleven prominent (male) astronomers were asked to name the most important American astronomers then living. The top fifty astronomers then living would receive a star in front of their entry in the first edition of *American Men of Science* [39]. Fleming was, on average, ranked 36th, validating the media's treatment of her as a leading astronomer. Pickering, however, ranked her 11th [40]. She had come a long way from 'copying numbers on prearranged forms'!

A sense of the work the women of the HCO performed, and the way in which they worked together, can be gleaned from another entry in Fleming's journal, describing her work during a single, not atypical, day (Fleming uses a passive construction for much of what she did, even in this journal form):

> Mr. Waite called early to see the Director and I had quite a talk with him about his studies and his prospects…Part of the morning I spent with Miss Cannon, discussing the remarks on her Classification [of stars] and explaining the reasons why we had changed 'one thing and questioned another'. Then Miss Leland was interviewed regarding her selection and measurement of the 'Faint Stars for Standards of Stellar Magnitudes'. This was followed by an interview with Miss Mabel Stevens relative to the checking of the identifications of these same stars in the Durchmustering Catalogue. Before lunch I found time to examine a few southern spectrum plates and marked a fourth type star and a gaseous nebula, both probably known. Later in the afternoon I noted a few more interesting objects, among them two fourth type stars, one gaseous nebula, and several bright line stars. Some of these may be new.

Fleming continues with a lament familiar to mid-career scientists:

> Looking after the numerous pieces of routine work which have to be kept progressing, searching for confirmation of objects discovered elsewhere, attending to scientific correspondence, getting material in form for publication, etc., has consumed so much of my time during the past few years that little is left for the particular investigations in which I am particularly interested…I have delegated my measures of variables, etc., to Miss Leland and Miss Breslin. I hope, however, to be able soon to finish the measures of the 'out of focus' plates, and to get well settled down to my general classification of faint spectra for the New Draper Catalogue.

After Fleming's death in 1911, Pickering tried to correct the published record. He issued an index of the *Annals* which listed authorship for each paper and volume. In addition to crediting Fleming with her early, anonymous papers, the Draper Catalogue, which originally had listed only Pickering's name on the title page, was now listed with only Fleming's[5].

---

[5] While the clear intent of Pickering in issuing the index was that authorship of the Draper Catalogue should be given to Fleming, his attempt did not succeed. Most modern databases still list Pickering as the author, a convention I therefore reluctantly follow in the citation list for this chapter.

As Pickering's own career advanced, he became a firm convert to the cause of women practicing astronomy. From 1914 to 1919, he arranged for a fellowship to be offered to women wishing to conduct astronomical studies [41], contributing his own money in support. In 1916, donors endowed a second fund on the occasion of Pickering's fortieth anniversary as Director of the Observatory [42], to be named after him.

In 1919, Pickering was working on an announcement regarding that year's fellowships for the *Harvard College Observatory Circular* when he fell ill, dying a few days later. Annie Jump Cannon, one of the great astronomers who had developed under his guidance, finished the article.

Four years later, in 1923, the Pickering Fellowship was awarded to an English student of physics named Cecilia Payne.

2.4 Starstuff

Shapley, at first, did not know quite what to make of this new addition to the HCO. Payne has said that, when she triumphantly wrote to Shapley to say that she had raised the money and was coming to Massachusetts, he did not seem to remember her [43], despite considerable correspondence with and about her predating her arrival [44].

Since she was supported by fellowships and came highly recommended, however, Shapley was happy enough to have her. He suggested she work in photometry, i.e., in determining the apparent brightness of stars observed at the HCO. This was crucial work, and had been a focus of the HCO for decades, but would have made little use of Payne's training in physics. Instead, Payne drew on her work with the astrophysicist Eddington and her friendship with his assistant E A Milne. Milne had become fascinated with the work of Meghnad Saha, of Allahabad University in India, on the theoretical interpretation of stellar spectra [45]. With Fowler, another acquaintance of Payne's, Milne wrote a seminal paper refining Saha's work [46]. Payne proposed using their results to analyze data from the library of stellar spectra collected at HCO, with the hopes of using them to determine the relative abundances of each element present in the stars. Shapley did not hesitate, directing her to the HCO's collection of more than a quarter million stellar spectra recorded on photographic plates (figure 2.5) [43].

This, however, created an immediate problem. While the project Payne proposed to do was important, timely, and a good use of her talents, she was not the first to begin work on it. Princeton graduate student Donald H Menzel, under the guidance of the renowned Henry Norris Russell, was pursuing similar studies for his dissertation, using, with Shapley's permission, the same HCO plates to which Shapley had just directed Payne! Shapley's solution was, with Russell's consent, to try to divide the problem between them, directing Payne and Menzel to study different types of stars [47].

Both Payne and Menzel thought of their work as astro*physics*, and not just astronomy, and thus were interested in testing and applying the physical theory of Saha, Milne, and Fowler over as wide a range of circumstances as practicable. While dividing data between them would have made sense if they were pursuing the kind of

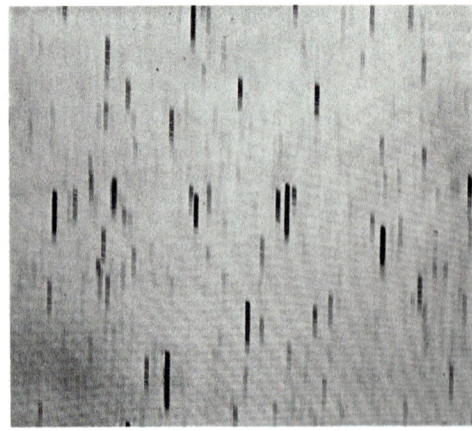

Figure 2.5. One of the HCO plates. Each vertical smear is the spectrum of a star. With permission from Harvard College Observatory.

observational classification which had been the HCO's bread and butter for decades, in this case it was more like Tycho Brahe's decision, some three centuries prior, to provide Johannes Kepler data only on Mars as Kepler tried to work out the laws that governed planetary motion. Payne eventually came to see Shapley's decision to separate their work in that way as 'divisive' and 'a great opportunity missed' [1].

At first, Payne worked largely on her own. This was partially due to Shapley discouraging her from direct collaboration with Menzel, but was in accord with Payne's own wishes at the time:

> I used to think: 'This is *my* problem'. I guarded it jealously; I snarled at anyone who dared approach it [1].

Payne immersed herself in this solitary project, working long hours. In fact, she later thought she might have contributed to stories of the ghost of Henrietta Leavitt haunting the HCO, for Payne now worked at Leavitt's old desk, often into the wee hours of the night. Shapley was supportive, but eventually asked if she thought she should perhaps publish something. She replied that she would feel she had failed if she published before solving the problem. While in her autobiography Payne suggests that Shapley was pleased by that response, it is not unlikely that he was simply content to allow her to conduct her research as she saw fit, both since she was on a fellowship and was not an employee under his direction, and because of the complications with Menzel and Russell.

She eventually sorted out the contribution of silicon to stellar spectra, using the theory to estimate the surface temperature of hotter stars. She considered publishing as the androgynous 'C H Payne', but Shapley convinced her otherwise, and the paper appeared using her full first name [48], as had also been true of the paper she published while at Cambridge.

Less than a year after Payne came to the HCO, Menzel finished his dissertation. As it turned out, Payne and Shapley need not have worried. While Menzel went on to a successful career as an astronomer, his dissertation was hardly a tour de force, consisting of twenty rambling pages, including a one-page summary, a six-page data table, and a citation to Payne's work on silicon. As for the Saha–Fowler–Milne theory he was purportedly testing, he felt that the 'qualitative agreement [with the data he examined] is excellent', but that the 'quantitative agreement of observation and theory is, however, not so satisfactory'. Menzel then went on to admit that even qualitatively there were disagreements, such as 'the unexplained behavior of hydrogen' in certain stars. In conclusion, Menzel suggests that 'some revision of theory is necessary' [49].

Russell's biographer David DeVorkin argues that Russell himself was avoiding facing the facts, an attitude which may have then influenced Menzel. Spectroscopy had opened up a means of finding out what stars were made of, and the results had, at first, seemed to be a dramatic confirmation of what astronomers call the 'Copernican principle', the notion that there is nothing special about the Earth. Copernicus himself removed the Earth from its *position* at the center of the Universe; observations as far back as Galileo suggested that there was nothing special about its *material* as well. This supposition received dramatic confirmation in 1862, when Gustav Kirchhoff demonstrated that many of the lines present in the solar spectrum corresponded to elements found on Earth [50]. By 1901, the American physicist Henry Rowland felt that, 'were the whole Earth raised to the temperature of the Sun, its spectrum would probably resemble that of the Sun very closely' [51], a sentiment which Russell echoed in a paper he published in *Science* in 1914 [52]. This assumption, while not assumed to be rigorously and precisely true, guided the development of theoretical models of stellar interiors by Eddington and, in turn, Russell's own theories of stellar evolution.

Eddington's models depended on the average molecular mass of particles in the stellar interior. This would seem to have required a detailed knowledge of chemical composition, but Eddington realized that deep within the stellar material, atoms would be nearly completely ionized, thus providing the same average molecular mass in a wide variety of cases. For example, suppose the Sun were made entirely of fully-ionized iron. Since iron has an atomic number (nuclear charge) of 26, there would be 26 ionized electrons for every one nucleus. For iron, the nucleus has a weight of 56 amu; each electron has a negligibly small mass in comparison. Thus, the average particle mass would be 56 amu divided by the 27 particles, or 2.1 amu. If, instead, the Sun were made of carbon, the same calculation would yield 1.7 amu; if it were made of lead, the result would be 2.5 amu. To the extent that atoms were incompletely ionized, these numbers would be a bit larger. Eddington knew that he couldn't pin down the precise value. In fact, when he made a computational error unrelated to this assumption, he simply changed his assumption from 2.0 to 2.8 amu to cancel out his other error and avoid having to go back and redo the calculations [53]! If the Sun were mostly hydrogen, though, that would be different; the average atomic mass would then plummet to 0.5 amu. Eddington was aware of that possibility, and explicitly acknowledged it in his discussions of his models. But if that possibility turned out to be true, he would have to rework all of his calculations, an extraordinarily arduous prospect. Since Russell's theories were based in part on these models, Russell, too, had a vested interest in hydrogen being only a minor component of the Sun.

Nevertheless, the observational evidence for a large hydrogen abundance was clear. Spectral lines of hydrogen showed great strength over a wide range of stellar spectral types, and indeed the strength of hydrogen lines formed the basis for Annie Jump Cannon's widely adopted system of spectroscopic classification of stars. The strength of the lines also appeared to depend strongly on pressure, which would not be expected unless the abundance was very high. One proposed solution, occasionally brought forward by Eddington, was that the *atmosphere* of stars was rich in hydrogen, but that the interior was not [54]. Russell, in collaboration with the American physicist Karl Compton (not to be confused with his more famous brother Arthur), took a different approach, arguing that hydrogen had 'special' electronic properties, perhaps shared in part by helium. In a paper they published in *Nature* in 1924, they wrote that failing to presume those special properties would 'demand an absurdly great abundance of hydrogen' [55]. It is little wonder than his student Menzel skirted around the issue.

In any case, with Menzel's dissertation finished, the path was cleared for Payne. As she doggedly solved the puzzle of element after element, publishing along the way [56], Shapley's enthusiasm grew. At one point, when Payne decided to attempt inverting the usual application of her technique to stellar spectra, using the surface temperature of a star to determine ionization potentials for species that had not been determined in a terrestrial laboratory, Shapley insisted she write it up for publication [57] …in a journal with a deadline the next day! While Payne inverted the application of her technique in the copy she wrote out longhand, working far into the night, Shapley inverted the traditional gender roles of the time, typing up each page for Payne as she went [1].

Shapley now felt that Payne should pursue a PhD in Astronomy at Harvard, but found that this prospect faced several barriers:

1. Payne was a woman, and Harvard did not give degrees to women
2. Payne did not, initially, want to get a doctorate
3. Harvard did not have an astronomy department

Nevertheless, Shapley persisted. The problem of Harvard being male-only had been addressed, albeit imperfectly, once Radcliffe, the sister school to Harvard, began issuing doctorates in 1902 [58]. Already, another Pickering Fellow, Adelaide Ames, had received a master's degree in astronomy from Radcliffe [59].

Shapley's persistence overcame Payne's reluctance, which was probably never particularly deeply held, solving the second problem. This left the third: while research in astronomy had a long tradition at Harvard, teaching had been neglected, and thus there was no department, either at Harvard or at Radcliffe, which offered a doctorate in astronomy.

Shapley, however, kept at it, securing approval from the chair of the physics department at Harvard and the authorities at Radcliffe [44].

In so doing, Payne's project became her dissertation topic, and the astronomy department at Harvard was born.

Figure 2.6. The 32nd Meeting of the AAS. Attendees mentioned in this text ('left' and 'right' refer to from the point of view of the camera, not the participants): Payne, appearing somewhat ghostly in a white dress, is the woman right of center toward the back. Shapley is immediately to the front and right of her in a dark suit, staring intently at the camera. Russell, a row or two in front of Payne and a bit to the left, is, in contrast, not looking at the camera at all. Cannon is the woman in the floral dress in the center of the front row, with Eddington to the left of her and Young just behind and to the right of her. Maury is the hatless woman a few people to the left of Russell. With permission of American Astronomical Society. https://had.aas.org/resources/aashistory/early-meetings/1922-1927#32.

Russell, meanwhile, with Shapley's encouragement, turned his attention to providing guidance for Payne. In mid-June of 1924, shortly after the completion of Menzel's dissertation, he characterized Payne's nascent dissertation as 'a first class piece of work'.

In August, Payne, as part of a contingent from HCO including Shapley and Cannon, attended the 32nd meeting of the American Astronomical Society (AAS) at Dartmouth (figure 2.6). Russell and Eddington were also in attendance. While there, she was elected to membership in the AAS, the organization that resulted from the 1898 meeting at which Pickering read Fleming's paper. It was, by this time, relatively welcoming to women working in astronomy. Among the officers of the organization, for example, was Dr Anne S Young of Mount Holyoke, who had received her own PhD in astronomy at Columbia [60][6] nearly twenty years before [61]. In all, at least a dozen of the 68 astronomers present were women [62]. Payne was in the vanguard of women in astronomy, but she was far from alone.

---

[6] Columbia, unlike Harvard, had been granting occasional doctorates to women for decades, beginning with the astronomer and mathematician Winifred Edgerton in 1886.

That fall, Russell visited her twice at Harvard to discuss her work. In contrast to Menzel, she did not limit the scope of her investigations or conclusions without a fight. Russell's first visit left her, in her words, 'in a state of prostration'. His second visit, a month later, left them *both* that way [47]!

In January of 1925 Russell responded to the latest version of her work, revised after their fall meetings. He was generally favorable, but provided two sets of suggested corrections: one for magnesium, magnesium (I) ions, and potassium, and the other for hydrogen, helium, and oxygen.

By this time, Payne's computations explicitly suggested that the surface layers of the Sun were predominately hydrogen, a conclusion Russell was not ready to accept. 'I am convinced that there is something seriously wrong with the present theory', he wrote in a letter to Payne. 'It is clearly impossible that hydrogen should be a million times more abundant than the metals...' [63].

When Payne published her dissertation later that year, she cited Compton and Russell's 1924 paper and accepted its premise that hydrogen abundance in the Sun could not possibly be as large as a straightforward interpretation of the data suggested, while maintaining caution about the particular form of their explanation, as it did not explain how helium also appeared to be superabundant in stellar spectra. In puzzling over this problem, Payne was reduced to quoting Russell's letter to her directly in the dissertation:

There seems to be a real tendency for lines, for which both the ionization and excitation potentials are large [that is, for atoms such as hydrogen and helium], to be much stronger than the elementary theory would indicate [64].

She again quotes a letter from Russell to her directly, this time in a footnote, regarding a discrepancy in iron between her computed stellar abundance and its expected abundance in the Earth:

Professor Russell believes that iron is much more abundant, at least in the Sun, than calculated above. He writes...

To me, these instances where she references Russell via a direct quote suggests that she wishes to keep her distance from their content, particularly when introduced by the relatively weak phrase 'Professor Russell believes' rather than, for example, 'Professor Russell has provided an explanation'.

Payne shows little of that hesitancy, however, when it comes to the crucial question of the *abundances* of hydrogen and helium (as opposed to the related question of *why* their lines appeared to be so strong):

The outstanding discrepancies between the astrophysical and terrestrial abundances are displayed for hydrogen and helium. The enormous abundance derived for these elements in the stellar atmosphere is almost certainly not real. Probably the result may be considered, for hydrogen, as another aspect of its abnormal behavior, already alluded to; and helium, which has some features of astrophysical behavior in common with hydrogen, possibly deviates for similar reasons. The lines of both atoms appear to be far more persistent, at high and low temperatures, than those of any other element.

What was she thinking as she wrote that text? Did she, as Neil deGrasse Tyson characterized it [65], 'cave' to Russell's authority? Was this, as Russell's biographer DeVorkin suggests, a canny attempt by Payne to, by denying the evidence of her own data, get her analysis showing very high abundances for hydrogen and helium published right under Russell's nose and with his approval? Or did she truly disbelieve her own results?

Payne's own recollections do not clarify the matter. When, decades later, Payne was asked about it in an interview by the astronomer and historian of science Owen Gingerich, she claimed not even to remember what she had written: 'Oh, did I say that? Well, pretty soon I convinced myself that [the high abundance was real]. I would have said that I always thought that it was so. But it certainly didn't take me very long to be convinced that it was' [43].

Gingerich reported that in another conversation she said that 'probably Henry Norris Russell talked me out of it' [44].

Payne was more consistent, however, in her recollection of an incident that took place just a few months after her dissertation was published. Visiting her idol Eddington in Cambridge, she said that she had discovered that there was more hydrogen in stars than anything else, to which Eddington replied, 'you don't mean *in* the stars, you mean *on* the stars', reflecting the theory that the stellar atmospheres were mostly hydrogen, but not necessarily the interiors[7].

In contrast to this recollection, however, stands the text of a short radio segment Payne broadcast on 8 December, 1925. In it, she says:

There is, however, one star of which we can very easily obtain a piece, though too large a piece for convenient analysis. Because the planets were born from the atmosphere of the Sun, the Earth is a good sample of the building materials of the stars. It is difficult—indeed it is impossible—to analyze the Earth. At best we can examine only a thin layer at the surface, knowing that the composition must be different lower down. But it can be said fairly definitely that the seven commonest elements of which Earth is made—oxygen, iron, silicon, magnesium, aluminum, calcium, and sodium—are also the commonest constituents of the stars' [66].

More than six months after finishing her dissertation, Payne was stating with confidence that hydrogen and helium were *not* among the most common constituents of stars.

---

[7] Payne's recollection of the precise phrasing of this exchange is different in [1] than [43], but the gist is the same.

While the question of Payne's beliefs and intent while writing her dissertation may never be conclusively resolved, in the following pages I will discuss several possible explanations, and how I've come to my own tentative conclusion.

Explanation 1: Payne deferred to Russell's expertise, assuming that he knew astronomy better than she.

If we knew nothing about the rest of Payne's life and writings, this would be plausible. Russell was *the* authority in American astronomy at the time, and Payne a mere graduate student…not to mention the gender dynamics that would have been present.

But Payne was not one to defer to authorities lightly. She clearly held her own, for instance, in her contentious meetings with Russell in the fall of 1924.

Perhaps most tellingly, in her autobiography she tells a story about a scientific matter on which she *did* defer to the authority of Russell and Shapley:

When she was still an undergraduate at Cambridge, she had come across an astronomical problem that vexed her. She believed a spectral shift known as the Stark effect should be seen in some stellar spectra, but it was not discussed in the astronomical texts of the time. Riding her bicycle up to the Solar Physics Observatory, she accosted the first person she saw with her question. The man turned out to be Milne of the Saha–Fowler–Milne theory. Milne admitted that he didn't know the answer to her perceptive question, and they soon became good friends.

After arriving at the HCO, she tried again, this time finding data that suggested the Stark effect was visible in some spectra. She brought the idea to Shapley, who shared it with Russell. Both were skeptical, and she was convinced not to publish [1]. Nevertheless, Payne did not back down completely, including a carefully cited discussion of the proposed effect in her dissertation and suggesting that it was important to keep the 'possibility in mind'.

Five years later, experimental evidence for the effect was independently discovered and published [67]. Decades later, Payne was still kicking herself over (partially) backing down:

> I was to blame for not having pressed my point. I had given in to Authority when I believed I was right. That is another example of How Not To Do Research. I note it here as a warning to the young. If you are sure of your facts, you should defend your position [1].

This missed opportunity occurred during the same period of her life as the question of hydrogen and helium abundance, with the same principals (Payne, Shapley, Russell), involving the same kind of data. In fact, it could so easily have been applied to what she wrote about hydrogen and helium abundance in her dissertation that some modern accounts use the quote out of context, implying that it applies to her dissertation.

I expect that if Payne had felt that she had caved on the question of hydrogen and helium abundance in her dissertation, she would have included it in her regrets, just as she did with the stellar Stark effect.

Explanation 2: Payne was misled by Russell, who wanted to steal her idea.

This explanation is included for completeness, as it's alive and well in some of the less reliable corners of the internet. While such things have happened all too often in the history of academia, and continue to happen today, this is not one of those cases. Russell's personal incentive here would be to bury the idea, as it would undermine much of his other work, not steal it!

Explanation 3: Payne was misled by Russell, who wanted to suppress her idea.

Unlike explanation 2, this is not completely implausible, and may have contributed to what transpired. It's quite possible that while this was not a *conscious* motivation on Russell's part, his desire to save Eddington's models and his own theory of stellar evolution caused him to be more dismissive of Payne's results than would otherwise have been the case.

Explanation 4: Payne, educated as an astronomer in the early 1920s, believed, as did most astronomers, that the composition of the Sun was similar to the composition of the Earth, and thus doubted her own results.

This view is summed up by Gingerich:

> She really didn't know that she had it, that that is the right abundance of hydrogen. It is very difficult, I think, and my reading of it is that she was fully accepting the arguments against it and therefore didn't have that as her discovery even though she had made those calculations in her thesis...Because everybody at that time believed in the uniformity of the Universe and therefore assumed that the Sun should be similar in composition to the Earth and so this was the basis of Eddington's making stars primarily out of iron and so on [68].

But why, if she were already thinking this way, did Russell need to repeatedly emphasize to Payne, as she was writing her dissertation, that her results for hydrogen and helium were wrong?

Explanation 5: Russell was tricked by Payne, who wanted to get her analysis in to the scientific record without running afoul of Russell.

This is the idea suggested by DeVorkin. But if this was her plan, why did she not at least mention hydrogen in her radio address later the same year?

In addition, Payne was not shy, in later years, about admitting other 'schemes' that she had employed. For example, she knew that the 1924 topic of the prestigious Adams Prize at Cambridge had been chosen so that Fowler, of the Saha–Fowler–Milne theory, was certain to win. And if Cambridge would not award women a full degree, it certainly wasn't about to award one a prize of this kind, and in fact did not do so until 2002. So, to show that she belonged in the same conversation, Payne gave her dissertation a subtitle based on the prize topic chosen for Fowler.

Payne also had a well-formed opinion of Russell's personal traits: he was 'selfish, overbearing, opinionated and conceited'. She said that 'you could admire him but I never could like him. He sort of gave me the creeps'. And:

> one of the things that was alarming was the enormous amount of power he wielded. Fortunately for me, he backed me, but if he hadn't...[43]

If Payne had actually pulled a fast one on Russell, wouldn't she have told of her victory in the years after he passed away?

Explanation 6: Russell believed he was right, and (mostly) convinced Payne.

This explanation assumes everyone was acting in good faith, and Russell simply won the scientific argument. As is often the case early in scientific disputes, the *truth* may not yet have had the preponderance of *evidence* on its side. There was a good deal of astrophysical theory that had been built up on the assumption that the composition of the Sun was similar to that of the Earth, and those theories had achieved notable successes when compared to observation. A single line in a single letter did not convince Payne; it took months of exchanging drafts and letters, as well as face-to-face meetings, before Payne was finally convinced. Even then, her conviction was not deeply rooted, and soon dissipated, but at least while she was writing her dissertation, it was sincere.

That is not to say that scientific conversions never happen all at once, via a letter. In fact, Payne was present when that happened to Shapley. For years, Shapley had been arguing that our Galaxy was the only one of its kind, and that the spiral nebulae viewed through telescopes lay within it. In late 1923, Hubble had discovered a Cepheid variable within the Great Spiral Nebula of Andromeda and then, using a relationship developed by Henrietta Leavitt at Harvard, determined its distance to be 300 000 parsecs, thus placing the Andromeda Nebula outside of, and equal in stature to, our own Milky Way Galaxy. The Andromeda Nebula had become the Andromeda Galaxy. In early 1924, Hubble sent a letter detailing his results to Shapley [69].

Payne was in Shapley's office when he read it. 'Here', he said, holding the letter out to Payne, 'is the letter that has destroyed my universe' [1]. From that point forward, Shapley dedicated himself to studying these other galaxies; his conversion had taken place in the time it took him to read the letter.

Again, if Payne had been particularly shaken by a *single* letter from Russell, wouldn't she have drawn the parallel to Shapley's letter from Hubble?

My thoughts on the matter

There does not have to be a single answer. People often act from a variety of motives, and in such cases may have a particularly hard time reconstructing those motives later. The way Payne discusses these aspects in her dissertation, her lack of a definite recollection of what she had written, and the text of her radio address, together suggest to me that she did not feel an active conflict at the time; i.e., that she managed to convince herself that what she was writing was scientifically defensible. Her lack of rancor toward Russell over this sequence of events, despite her evident distaste for his personal style in scientific matters, suggests that she does not feel she was misled or successfully browbeaten. On the other hand, when she first computed stellar abundances she may have taken them at face value, thinking they showed a large abundance of hydrogen and helium, and within a year or two of completing the dissertation may have reverted to that view.

Russell, for his part, was likely influenced, to some degree subconsciously, by his stake in Eddington's models, which in turn depended on stars being made primarily of anything but hydrogen. But if so, then he himself held a deep belief that the hydrogen abundance in stars was not high. He would not have been 'tricking' Payne, except to the extent that he was tricking himself.

Of course, more cynical interpretations are possible, particularly if we allow for the possibility that Payne, over her career, deliberately covered up her motivations and/or Russell's actions at the time. Payne does not strike me as that kind of person, but I leave it to my readers to form their own opinions.

2.5 Two astronomers from Cambridge

Even aside from the issue of the abundance of hydrogen and helium, Payne's dissertation is a tour de force. Fifteen chapters, five appendices, and two indexes carefully cover all aspects of the relevant theories, including their limitations and complications.

Like Newton's *Principia*, critics could dispute Payne's interpretations or her conclusions, but could not gainsay the evident skill and knowledge of the author.

In fact, while the scope and impact of Payne's dissertation is necessarily more limited than *Principia*, the works are organized in a similar fashion. This may not have been an accident. Payne first read the *Principia*, in the original Latin, at the age of 12 [1]. She later attended Cambridge, as had Newton. Newton eventually became a professor at Cambridge, and would have been thought of as a kind of patron saint of both mathematics and physics there. Surely, every student of physics to pass through its gates must have seen themselves as responsible for Newton's legacy. When it came time to write her dissertation, Payne may, whether consciously or not, have chosen to use a structure similar to Newton's.

Like the *Principia*, Payne's dissertation is divided into three large parts. In both works, the first two parts use a combination of theory and observation to establish the analysis on a firm footing. In, the third section of each, the tools from the previous sections are used to draw sweeping conclusions about the Universe. Finally, thoughts are given to future directions for study.

Related to the structure of each work was its purpose. While popular culture attributes to Newton the idea that it is gravity that makes the planets go around the Sun, the Moon around the Earth, and the apple fall from the tree, the idea had been raised before. It was, so to speak, 'in the air'. Even the gravitational inverse square law had been anticipated by others, as Newton himself acknowledges in the *Principia's* Book I, Proposition 4, Corollary 6. The power of the *Principia* is not so much that it introduces new ideas, but that it creates a coherent 'system of the world' which brings those ideas together.

Payne's dissertation is equally clear about what it has accomplished. The conclusion to the chapter on elemental abundances begins with a definitive statement that the surface layers of all stars are made of the same stuff, in roughly the same proportions: 'The uniformity of composition of stellar atmospheres appears to be an established fact'. The passive voice obscures her meaning: it was Payne, through her dissertation, who promoted that idea from a common, but unproven, supposition to a known fact.

In the next paragraph she argues that the uniformity applies not just to the surface layers, but to the stars as a whole, extending the scope of her claim.

Following the chapter on elemental abundance is a chapter boldly titled 'The Meaning of Stellar Classification'. Payne discusses at length the empirical (as opposed to theoretical) origins of Cannon's classification system, and then provides this remarkable passage:

> Although devised with no theoretical basis, ...[Cannon's] classification has long been recognized as classifying something physical, and the fact that the majority of the stars had been ranged by it in a single sequence suggested that a single variable was principally involved. From general theoretical considerations it could have been predicted that this variable was probably the temperature, but, in addition, the observational evidence that this was the case was immediately convincing...The preceding eight chapters review the arguments and the observations that have established the connection between the spectrum of a star and its temperature. From an examination of the data there given it becomes clear that what ...[Cannon's] system classifies is essentially the degree of thermal ionization.

'Degree of thermal ionization' is not quite the same thing as temperature, because it depends on pressure as well. Thus, two stars with the same degree of thermal ionization, but with different atmospheric pressures in the regions that produce the line spectra (as is the case for giant and dwarf stars), would have slightly *different* temperatures, despite belonging to the same spectral class. This fine-tuning of the meaning of spectral classifications, while difficult to encapsulate in a sound bite for the general public, was crucial to practicing astronomers.

Again, a comparison to Newton might be instructive. By Newton's time most, but not all, astronomers were becoming convinced that the Sun, rather than the Earth, was the center of the known Universe, with the Earth orbiting around it. Newton's thorough work removed all doubt that the Earth was not the center, but, in what amounts simultaneously to a technical correction for practical purposes and a seismic shift for philosophical ones, neither was the Sun:

> Hypothesis I: That the center of the system of the world is immovable. This is acknowledged by all, while some contend that the Earth, others that the Sun, is fixed in that center. Let us see what may from hence follow.

Proposition/Theorem 11: That the common center of gravity of the Earth, the Sun, and all the planets, is immovable...

Proposition/Theorem 12: That the Sun is agitated by a perpetual motion, but never recedes far from the common center of gravity of all the planets...hence the common center of gravity of the Earth, the Sun, and all the planets, is to be esteemed the center of the world...if that body were to be placed in the center, toward which other bodies gravitate most (according to common opinion), that privilege ought to be allowed to the Sun; but...the Sun itself is moved...

In short, Newton suggests that for 'common opinion' it is OK to say that the Sun is the center of the solar system, but in actuality the Sun itself moves around a bit, jostled from place to place by the gravity of the planets. The true, and stationary, center is the center of gravity of the whole system. This is analogous to Payne's distinction: it is reasonable to discuss temperature as the organizing principle of stellar spectra when conversing with lay people, but that is not strictly correct; the key variable is actually degree of thermal ionization, which also depends to some extent on pressure.

2.6 Reactions

Reaction to Payne's dissertation was uniformly positive. Russell wrote Shapley, saying that it was 'the best doctoral thesis I ever read', before diplomatically allowing an exception for Shapley's own, more pedestrian, dissertation [70]. Russell continued:

[It] ought to be strongly recommended, not only to observatory libraries, but to all students of the subject...I am especially impressed with the wide grasp of the subject, the clarity of the style, and the value of Miss Payne's own results [71].

Formal reviews followed. *Nature* called it 'an indispensable handbook' [72], and a review in *Publications of the Astronomical Society of the Pacific* stated:

...this book constitutes an important contribution to astrophysics. The doctors' degree will not fall into disrepute if this standard of excellence for the thesis be maintained. [sic] [73].

A reviewer in *Physical Review* went even further, gushing that 'the book is worthy of a place in every physical, as in every astronomical, library' [74].

Otto Struve, writing in the *Astrophysical Journal*, took his responsibilities as a reviewer seriously, drilling down in to such minutiae as a typo in one of the tables. His overall opinion was positive, if restrained:

Stellar Atmospheres while written in a popular and interesting style, differs from the conventional treatise on astronomy in that it is intended not so much for the layman as for the advanced student and the specialist. It is an important addition to American bibliography... [75].

Many years later, Struve's opinion of Payne's work had evidently increased, as he called it 'undoubtedly the most brilliant PhD thesis ever written in astronomy' [76].

It is worth noting that none of these reviews made an issue of Payne's gender; the focus was almost entirely on her science and the clarity of her writing.

Eddington's reaction, in contrast, was among the most peculiar in print. He clearly thought well of the work, citing it repeatedly in his 1926 treatise, *The Internal Constitution of the Stars* [77]. But when it came to the question of elemental abundances, a question key to his models, his use of her work runs off the rails. He first described her method, judging it 'not so wild as we might suppose at first', a characterization which apparently delighted Payne [1, 43]. He then reproduced her final table of results, from which she omitted hydrogen and helium, since she believed (or, at any rate wrote), that the high abundance implied by her calculations was spurious. So far, so good. But then he provided this interpretation:

> Other elements which are probably abundant are O, S, N, Ni, but quantitative determination is not yet possible. Information is not obtainable as to P, Cl, F, Zr which are terrestrially abundant. Miss Payne considers that there is a fairly close parallelism shown between stellar abundance and terrestrial abundance.
>
> A study of this table does not suggest any need for amending the view...that the mean molecular weight should be taken to correspond to a predominance of elements in the neighborhood of Fe with some admixture of lighter elements.

In this quote, Eddington provides a long list of elements which were not in Payne's final table for a variety of reasons, but omits hydrogen and helium. He then claims his assumption of a mean molecular weight of around two (2.1, in this particular book) is justified by her work! Scientific integrity should have required Eddington to acknowledge that the large amounts of hydrogen implied by Payne's analysis would, if true, undermine his models. The fact that Payne herself dismissed them does not, to my mind, free him from that responsibility, and certainly suggests that he should not have used her abundances as *evidence* for a mean molecular weight consistent with a low hydrogen and helium abundance.

2.7 Blocked paths

Following the publication of her dissertation, Payne was famous, at least among astronomers.

In 1927, she received a signal honor, when the fourth edition of *American Men of Science* [78] came out. Payne was among the 250 scientists from across all fields of study who were added to the list of 'leading scientific workers', which had last been updated six years prior. At the age of 27, she was now one of the top astronomers in the United States.

If she had been male, it was not unlikely that this would have led to a prominent position at a university or observatory. But while the scientific community was willing to accept her and other women as equals in the world of ideas (as Payne put it, the words scientist and scholar do not have a gender), the same did not apply to employment. The historical record is littered with anecdotes of Payne being thought worthy of prestigious positions, but 'ineligible' because of her gender.

Russell, for example, in 1934 was considering who might eventually replace him as Director of the Princeton University Observatory. The best American for the job, wrote Russell, 'alas, is a woman!,—not at present on our staff'. It would be decades before Princeton allowed women to hold prestige positions [79].

Closer to home, there was the Department of Astronomy at Harvard, a department in effect created by Payne's doctorate. Payne was acting as an informal dissertation advisor to its first doctoral student, Frank Hogg, and also teaching classes to the new mixed-gender batch of doctoral students[8]. One of those new students, Helen Sawyer[9], later wrote that 'the combination of two brilliant and dynamic individuals, Harlow Shapely and Cecilia Payne, was the spark which brought the graduate school in astronomy at Radcliffe and Harvard into existence' [80]. It would have been natural for Payne to become its new chair, and she expressed an interest in the job:

The new Department called for a Chairman, a Professor. I could have done it; who knew the ropes better? But it was 'impossible'; the University would never permit it [1].

Instead, Harry Plaskett was brought in to fill the job[10], and Payne remained at the HCO as a 'Technical Assistant' to Shapley. Payne felt that, unlike other astronomers, Plaskett did not treat her as a fellow scientist. Just a few years later, in 1932, Plaskett landed a prestigious professorship at Oxford. 'Not for the first time', wrote Payne, 'I felt I had been passed over because I was a woman'.

Nor would it be the last. When Plaskett left, Shapley wanted to bring in a spectroscopist to chair the department. This was a double blow: Payne was, by training, a spectroscopist. But her job as Technical Assistant did not allow her the freedom to choose her own research direction, and Shapley had directed her away from spectroscopy and into photometry. Photometry had been Shapley's preference for Payne from the moment she first arrived, but at that time she had then been on a fellowship, and thus able to choose her area of study. Now, however, she was an

---

[8] The women would formally receive Radcliffe doctorates, as Payne did, while the men received doctorates from Harvard; the Harvard Astronomy department now, in retrospect, considers them all, including Payne, to be alums of the same department [81].

[9] While working on her doctorate, Frank Hogg and Helen Sawyer were married, at which point she changed her name to Helen Sawyer Hogg, the name under which she wrote much of the material cited here. To avoid confusion with her husband, or the need to provide first names, I will refer to her in this text by her maiden name, Sawyer.

[10] The position of chair was not made official until 1945 [82]. Plaskett and Menzel were *de facto* chairs during this period, even though they didn't have the title in official Harvard documents.

employee, a member of the staff. Payne argued her case, but without success. Instead Shapley invited Struve, the astronomer who had written the detailed review of Payne's dissertation, to be the new chair. When Struve realized this would mean that Payne would have to give up spectroscopy for good, he declined.

The next to be invited was Menzel—the same Menzel whom Payne had been competing against while they were graduate students. Unlike Struve, he accepted the position. Payne had decisively beaten Menzel in the battle of the dissertations, and had gained immediate renown in consequence. But in terms of titles, pay, and freedom to pursue his academic interests, there was no contest.

It is true that we don't know how this would have played out if gender had not been a factor. Sawyer, who, as one of the first of the new graduate students, would have had to finish her degree under Payne had she been appointed in place of Plaskett, had this to say:

> Yes, she blamed (and in the book too) her lack of promotion at Harvard on the fact that she was a women, not a man. My impression is, now I can't say how different it would have been had she been a man, but I think her disposition must have come into this because when I think back to the way she behaved in the late 1920s around that Observatory, I would shudder at the thought of her being the director of an observatory like Maria Mitchel or Mount Holyoke or whatever, even a small observatory and heaven forbid a big one, because you didn't know where you stood with her, she was so up and down and inflammable, you might say. As an observatory director, whether or not it was a man or a women behaving that way, you wouldn't want it, you need stability and in the 1920s Cecilia did not have it [68].

This, however, is a bit disingenuous of Sawyer. The Cecilia Payne of the 1920s was not vying for the position of the director of an observatory, but rather to be a professor and the chair of a department. Observatory directors have a staff and a budget; professors, even department chairs, have students and research. The issues of temperament that Sawyer identifies would be more injurious to performance in the former role than the latter. And how much of what concerns Sawyer is attributable to the fires of youth? Observatory directors were rarely younger than their mid-thirties when first stepping into the role. Even Russell's musings related to the possibility of grooming Payne to take over for him when he retired, by which time she would have been well into her forties.

We don't know, of course, how Payne would have fared if there were an even playing field. But to say the playing field was uneven is an understatement; Payne was not even allowed into the game.

2.8 Love (of science) levels all ranks

It was in this atmosphere, in 1929, that Shapley pulled out the manuscript for *The Harvard University Pinafore,* first written (perhaps by Mina Fleming!) in 1879. The play had not been performed when written, Shapley speculated, because it was 'a little too daring' [83]. But now, with another meeting of the AAS scheduled to take place at Harvard at the end of the year, perhaps the time was right.

Shapley summoned Helen Sawyer to his office, handed her the manuscript, and asked her opinion as to whether it could be produced. (Sawyer had experience with musicals from high school, and thus on this matter she was the resident expert.) She thought it could, at which point Shapley assigned her to choose Observatory staff for the roles and produce the play with him, now rechristened the *Observatory Pinafore* [21], with the goal of a New Year's Eve performance for the attendees of the AAS.

Josephine, the Captain's daughter of Gilbert and Sullivan's original, had, in the 1879 HCO version, become Joseph, a young and talented astronomer at HCO who has to decide between staying at Harvard or taking his talents elsewhere. In 1929, this role was given to Cecilia Payne, and Joseph once again became Josephine.

It is inconceivable that the HCO staff, especially Shapley and Payne, were oblivious to the resonances that this created.

The computers, female in 1929 as in 1879, still sung of their role:

> We labor hard all day;
> We add, subtract, multiply and divide,
> And we never have time to play.
> No, no; No, no,
> We never, never play.
> No, no; No, no,
> We never, never play.
> We sit at our desks all day, all day,
> We work from morn 'till night
> And computing is our duty,
> We are faithful and polite,
> And our record book's a beauty,
> Computing is our duty,
> Our record book's a beauty,
> We work from morn 'till night,
> We are faithful and polite.

The lot of astronomers, including 'Josephine' was also difficult, but in a different way:

> An astronomer is a sorry soul,
> As free as a caged bird;
> His sympathetic ear should be always quick to hear
> The directorial word.
> He must open the dome and turn the wheel
> And watch the stars with untiring zeal.
> He must toil at night though cold it be
> And he never should expect a decent salaree.

His eyes should shine with learned fire,
His brow with thought be furrowed;
His energetic speech should be ever prompt to teach
The truths which he has borrowed.

Whether Josephine should stay at Harvard or leave for greener pastures is fiercely debated during the play. Josephine's skill is repeatedly stressed, as is the fact that she has a degree that some of the other (male) astronomers lacked—all facts strangely well suited to Payne!

Only when one of the other astronomers threatens suicide if she leaves does Josephine agree to stay.

In the second act, one of the major themes from Gilbert and Sullivan's version is introduced, but with a twist: 'love levels all ranks', but here it is the scientific instruments and the scientific works, along with the people who use them, that inspire love. The 'ranks' in question remain, in the lyrics, those of social station. But it made little sense, either for Cecilia or for the Josephine portrayed in the play, that those ranks would be based on socioeconomic status. Instead, the stylized Victorian costumes the Observatory staff donned for the play accentuated a different kind of gulf between Josephine and the other astronomers (figure 2.7). The female computers wore white shirtwaist blouses and long skirts, belted by a sash featuring a stylized star or moon. The male astronomers, in contrast, wore dark suits. Payne, like the computers, wore skirt and blouse, perhaps in the same color palette as the computers (the only photographs of the play are in black and white, but the skirts of

Figure 2.7. Cast and crew of the *Observatory Pinafore*. Payne is seated in the front row, with a pair of prop prisms on her lap. Shapley is the short hatless man with the necktie behind and to the right of Payne. Sawyer is the woman at the right end of the front row, next to the kneeling boy in the necktie. With permission from Charles Reynes.

both computers and Payne register in the midtones). Distinct from the computers, however, she wears a jacket which matches her skirt, echoing the professional attire worn by the men. Unlike Payne's day-to-day attire which, in the words of her daughter, tended toward conservatism and even 'mannishness' [84], the costumes in the play brought the contrast in gender to the fore. The subtext would have been crystal clear, both to the participants and the audience.

The play ends with a bitter irony: in the final scene, Josephine is saved from disgrace when it is revealed that she has secretly been studying photometry. The astronomer who was trying to lure her away reacts in disgust:

Josephine will not do for my assistant. I cannot have anyone with me who is at all interested in photometry.

But the Director of the Observatory is delighted:

Josephine is evidently too valuable to the institution to be allowed to leave it.

Payne, of course, was by this time working in photometry because Shapley insisted she do so if she were to remain at Harvard. And yet, as the play suggests, other institutions were more interested in other research areas, such as the spectroscopy she'd been apparently forced to abandon.

It is simply coincidence that, in 1879, long before Payne was even born, a play was written that resonated so strongly with Payne's situation fifty years later. But it cannot be coincidence that Shapley, having discovered the manuscript eight years before, chose 1929 to stage it. It was, in his words, 'the appropriate time' [83].

In any event, the play was well-received at the AAS meeting. According to *The Journal of the Royal Astronomical Society of Canada*, '"the fair Josephine" (otherwise Cecilia H Payne) was easily the star of the performance' [85].

Whatever mixed feelings Payne may have had about the role of photometry in the play, she clearly found the overall experience invigorating, leading the assembled astronomers in singalongs of music from the play later that night and then again at dinner the next night [86].

And that was not the last performance of the *Observatory Pinafore*, as the cast reprised their performance for the community astronomy club Harvard sponsored. In their review, the Cambridge Chronicle noted that 'Cecelia [sic] H Payne, a woman astronomer of international fame, played the part of Josephine, the heroine, to the Queen's taste' [83].

As with Josephine, Payne remained trapped at Harvard, but as with Josephine, it was love of science that kept her there.

Eventually, bit by bit, change came.

On the scientific front, the truth about hydrogen was becoming clearer. By 1929 even Russell was convinced by the accumulating evidence. Following Sherlock Holmes' maxim that 'when you have eliminated the impossible, whatever remains, *however improbable*, must be the truth' [87], Russell now wrote that 'the obvious explanation—that hydrogen is far more abundant than the other elements—appears to be the only one' [88]. This quote appears near the start of Russell's extensive paper marshalling the available experimental evidence and theoretical tools to determine the composition of the Sun's atmosphere, and, by extension, that of all stars. Of Payne's dissertation, he writes that 'the most important previous determination of the abundance of the elements by astrophysical means is that by Miss Payne', which he describes as showing a 'most gratifying agreement' with Russell's more recent analysis using different theoretical tools and different data.

Using a tactic common in writing in the physical sciences (and not, generally, considered unethical), Russell makes it sound like the predominance of hydrogen should have been obvious all along:

> It is probable that the Earth and the meteorites were formed by condensation from matter ejected from the Sun...the ejected material must have been intensely hot, and would be likely to lose constituents of low atomic weight, hydrogen most of all...in the outer parts of the Sun, on the other hand, there are certain diffusional and electrostatic effects which tend to concentrate the hydrogen at the surface.

It is worth noting that, while acknowledging the accuracy of Payne's results, Russell has not entirely let go of the notion that the interior of stars might be different. Nevertheless, Payne's analysis was vindicated.

Should Russell have gone further, and incorporated an explicit *mea culpa* into his paper? It is difficult to see how he could have done so without implicating Payne as well. It would have been condescending and demeaning to suggest that she had published a dissertation that she did not believe, and saying so might have significantly damaged her career. Instead, Russell's paper implied that she made the same *interpretative* mistake that he did, but that her analysis was prescient.

I encourage those starting careers in science to ponder this sequence of events, for I believe there are no easy answers here. Should Payne have resisted the weight of scientific opinion in 1925, and the powerful arguments of Russell in particular, and defended her own results? Payne herself gave no indication that she harbored regrets regarding this, although she freely admitted mistakes she had made at other times. Should Russell have held back from arguing his point with her in 1925? That would have risked condescension; science is about a contest of ideas. Or should Russell have found a way to explain the sequence of events in his 1929 paper? That would have been a difficult needle to thread.

On the other hand, there was a way open to Russell that he did not pursue to any great degree, and for which I think he *can* be criticized. He could have followed his 1929 paper with public discussions of his interactions with Payne in 1924 and 1925, and credited her with glimpsing the truth four years before the evidence became overwhelming. That he did not do so is not, I think, to his credit.

As for Eddington, he reluctantly came to acknowledge that stars contained quite a bit of hydrogen, although he still favored values below 50%. In a 1932 paper in the *Monthly Notices of the Royal Astronomical Society* [89], he credited the discovery of high hydrogen abundance in stellar atmospheres to 'H N Russell and others', without explicitly mentioning Payne or even citing Russell. The lack of a citation to Russell is particularly galling, as it breaks the link back to Payne.

Finally, it is worth noting the citation to high hydrogen abundance provided by one other astronomer: Payne herself. In 1930, in a follow-up monograph to her dissertation which exceeded it in length, she stated that 'about 96 per cent of the atoms in the solar atmosphere are found by Russell to be hydrogen' [90] and cited Russell's paper, without discussing or citing her own 1925 computation, despite the fact that Russell includes that computation (with citation) in his own paper! It seems likely to me that Payne's own conceptions of scientific integrity, which could at times be unusually strict, prevented her from giving herself credit in that way[11].

It was through sloppy citation like that in Eddington's paper, perhaps combined with Payne's own reticence to claim credit for an idea she herself had rejected at the time, that Payne's role with regard to hydrogen gradually receded from awareness. As for Eddington's slight, it would have been less likely if Payne had been a professor in a leading department, as she clearly deserved, rather than a mere 'technical assistant'. In any case, a professorship would have kept Payne more in the eye of the scientific world at what should have been the height of her career. There were still those, including Russell, who recognized and acknowledged her brilliance, even if they weren't always shouting it from the rooftops. But as time went on, Payne risked fading from view.

That is not to say that Payne ceased being a productive astronomer, or even, despite Shapley's bidding that she focus on photometry, that she gave up spectroscopy. An examination of her publication list during her time as a technical assistant shows a roughly even split between works on photometry and those on spectroscopy. More often than not she was the sole author on pieces ranging from brief reports to lengthy articles, although her work also featured several different coauthors, including Shapley, Menzel, Hogg, and Russell. When she did have coauthors, she was lead author about half the time, suggesting she was not being passed over for that recognition because of her employment status.

It would certainly have been possible for her to drift away from the forefront of astronomy during this time. It is likely this period Payne had in mind when she gave this advice to Radcliffe undergraduates interested in becoming astronomers:

> Only become an astronomer if nothing else will satisfy you; for nothing else is (approximately) what you will receive. The material returns will never be great, and most of the scholars who receive them do so as the price of sacrificing time that they might devote to pure scholarship. Fame is the reward of the few, and those who achieve it realize too often that it has been purchased by efforts that were not their highest. If you do your job you will be able to earn your living, and you will taste the delights of discovery, than which (to the addict) there is nothing more satisfying. That is all. If it is enough, then become a scientist. If it is not, get out before it is too late [91].

---

[11] She had no hesitation, however, in citing herself in general, doing so more than two dozen times in the 1930 monograph.

Nevertheless, recognition of her continued efforts began to come in. In 1931, her senior colleague Annie Jump Cannon became the first woman to win the venerable Draper Medal for astronomical physics [92]. In 1932, Cannon followed this with the Ellen Richards Research Prize [16]. Cannon used her prize moneys to establish a new award, to be given once every three years to a woman who had made 'distinguished contributions to astronomy' [93]. In December 1934, Russell, at this point President of the AAS, presented the new award to Payne [94].

In 1938, with a new President at the helm of Harvard, both Cannon and Payne were finally given the formal title of Astronomer [95]. Payne was now a professor in all but name[12] and salary [1], with a professor's freedom to pursue projects of her choosing. Her publications from this period forward reveal a wide variety of interests and skills—she occasionally still worked in photometry or spectroscopy, but more often used both as tools for understanding novae, supernovae, and variable stars. At times she performed the painstaking exhaustive analyses which had been the hallmark of the HCO in Pickering's computer era, while in other papers she delved into the kind of astrophysical theorizing favored by Eddington and Russell. To these she added popularizations, opinion pieces, and reviews.

In 1950, Payne had a particular opportunity to show her range. The publication of Immanuel Velikovsky's pseudoscientific work *Worlds in Collision* was imminent, and the news magazine *The Republic* had asked Shapley to write a rebuttal of an article in *Harper's* [96], which described Velikovsky's theories [1].

Shapley passed the task on to Payne, who wrote an initial critique for *The Reporter*. The magazine made it clear that she was to evaluate it 'solely on the grounds of its astronomy', since *Harper's* had put forth the peculiar argument that the theory could be neither proved nor disproved because to do so would require 'a thorough knowledge of archaeology, paleontology, geology, astronomy, psychology, physics, chemistry, and several other sciences—as well as world history'.

For the first part of her critique, Payne limited herself to astronomy and physics, with a bit of the history of astronomy thrown in. But further into her review it seemed she could not help herself from discussing other nonsensical aspects of Velikovsky's theories, saying that geology, biology, classical literature, and Biblical scholarship were 'outside the province of the astronomer' but then commenting on those aspects anyway [97].

Payne's article ignited a firestorm of protest from Velikovsky and his supporters: she had not, they pointed out, even read the book! Payne therefore waited for it to come out, read it carefully, and produced a full critique for *Popular Astronomy* [98].

This time, she did not even give lip service to limiting herself to physics. In addition to demolishing Velikovsky's scientific arguments, she methodically dismantled his appeals to mythology and the Bible as well. This short excerpt will give a taste of the character of Payne's arguments:

---

[12] Since Menzel, at this time, was a Professor in title but a chair in all but name, the Harvard Astronomy Department was at this time suffering from a deficit of official status throughout.

Velikovsky informs us (p 81) that 'according to Herodotus, the final act of the fight between Zeus and Typhon took place at Lake Serbon', and refers us to Herodotus, III. 5. What does the Father of History actually say? On turning to the passage, we find: 'At this lake, where, as was reported, Typhon was concealed, Egypt commences'. The Greek word is the ordinary one for 'hidden'; nothing is said about Zeus or a battle.

Paragraph after paragraph, page after page, Payne relentlessly dismantled Velikovsky's work by the unusual approach of, for the span of the article, treating it as a serious attempt at scholarship. (Menzel, in contrast, published his own scathing commentary of Velikovsky, filled with satire and condescension [99]).

In an additional twist of the knife, Payne introduced the sections of her review with quotes from Shakespeare's *A Midsummer Night's Dream*; her section on Velikovsky's science began with a particularly straightforward line from the Bard of Avon: 'This is the silliest stuff that ever I heard.'

There were not many astronomers of the time with the knowledge and confidence to go directly at *both* the astronomical and mythological aspects of Velikovsky's work (although Russell certainly could have done it). Payne, given the chance, did not take half-measures.

At institutions like HCO, change often comes in waves, marking distinct eras. The departure of one Director and the arrival of a new one not only brings in new ideas and ways of doing things, but also allows long-simmering tensions to be resolved.

The early 50s were such a time for Harvard astronomy. Shapley had been Director of the HCO since 1921, and Menzel de facto chair since 1932. In 1952, after more than three decades, Shapley retired as Director of HCO, allowing Menzel to move into the role. For a time, Fred Whipple, another HCO astronomer with whom Payne sometimes collaborated, became chair of the department in Menzel's place[13].

When Menzel found out how little Payne was paid, he was dismayed. Despite being so concerned about the budget that he discontinued the collection of photographic plates of stellar spectra [100], he doubled Payne's salary [1].

Harvard as a whole was also entering a new era, appointing a new President. With all these changes, the stars were finally aligned; Cecilia Payne, in 1956, officially became a Professor at Harvard, with all the accompanying rights and responsibilities. But that was not all. Within a few months after her appointment, she was named Chair of the Department of Astronomy, becoming the first woman to chair a department at Harvard.

I ask myself what difference it has made to me as a scientist that I was born a woman. As concerns the intellectual side of the matter, I should say that it has made very little...on the material side, being a woman has been a great disadvantage. It is a tale of low salary, lack of status, slow advancement. But I

---

[13] Whipple became chair in 1949. Changes of leadership in powerful academic institutions are often orchestrated so as to occur in stages over the period of a few years, allowing for some continuity between eras.

have reached a height that I should never, in my wildest dreams, have predicted...it has been a case of survival, not of the fittest, but of the most doggedly persistent...I simply went on plodding, rewarded by the beauty of the scenery...young people, especially young women, often ask me for advice. Here it is, *valeat quantum*. Do not undertake a scientific career in quest of fame or money. There are easier and better ways to reach them...your reward will be the widening of the horizon as you climb. And if you achieve that reward you will ask no other.

—Cecilia Payne-Gaposchkin [1], shortly before her death at the age of 79.

2.9 Science summary: stellar spectra

Atoms and ions are made of a positively charged nucleus surrounded by negatively charged electrons. Atoms are neutral, and thus possess a number of electrons equal to the charge on the nucleus (measured in units of e, the magnitude of the charge on an electron). Thus an iron atom, with a nucleus of charge $+26e$, contains 26 electrons.

Ions are charged, meaning they have a different number of electrons than a neutral atom of the same element would. In stars, atoms tend to lose electrons, becoming positively charged ions. For example, if an atom of iron loses three electrons, leaving it with 23, it will have a net charge of $+3e$. Chemists refer to that iron as Fe^{3+}. Spectroscopists often use the notation Fe IV, because it is the fourth ionization state of iron, with neutral iron being the first. The labels in the spectroscopist's system and the chemist's system thus differ from each other by 1 unit. Astrophysicists tend to use the chemist's system when describing the physical state of the atoms in a star (e.g. there is a preponderance of Fe^{3+}) and the spectroscopist's system when describing spectra.

It's important to realize that the 'lost' electrons are still present in the star. Instead of being bound to an individual nucleus, they now wander freely throughout the region. The combination of free electrons and the remaining ions forms a 'plasma'.

For each atom or ion, the laws of quantum mechanics dictate that electrons are only allowed to have certain energies. These energies are usually specified relative to the energy of a free electron removed from the atom or ion. Since the atom loses energy when it binds to a nucleus (another way to think of this is that it takes energy to remove an electron from an atom or ion), energies expressed this way are negative. For example, neutral hydrogen's one electron can have an energy of -13.6 eV (eV is a unit of energy), -3.4 eV, or -1.5 eV. There are higher allowed energies (i.e. those closer to zero), but no allowed energies between -13.6 and -3.4 eV, or lower than -13.6 eV. The pattern of allowed energies for each ion and atom is different, and, with a few notable exceptions, they don't form simple mathematical patterns.

Suppose there is a hydrogen atom with its electron in the lowest possible energy state, -13.6 eV (the 'ground state'). If light, which is made up of photons, shines on the atom, it is possible for one of the photons to be absorbed, increasing the energy of the electron, but only if that would leave it in one of the allowed states. Thus, the atom could absorb a photon of energy 10.2 eV, bringing the electron up from −13.6 to −3.4 eV, which is allowed, but it could not absorb a photon of 10.0 or 10.4 eV, as that would not yield an allowed energy for the electron.

In the hurly-burly of a hot, *dense* plasma, atoms and ions rarely get a chance to execute such a simple transaction, as particles of all types are constantly colliding and exchanging energy. As a result, photons of all energies (meaning all colors) are emitted: a 'continuous spectrum'. Depending primarily on the temperature of the plasma, there may be more photons in one broad energy range than another, causing the dense plasma to appear colored to us: for example, depending on the mix, it may look red, or yellow, or a bluish-white. This is what gives stars their color.

But above the hot, dense plasma that provides us most of the light from stars, there are more rarified 'atmospheres'. Unlike on a rocky planet like Earth, there is no sharp dividing line between surface and atmosphere. Instead, the density of the plasma that makes up a star decreases as you get further from the center, eventually reaching densities much less than that of the air we breathe. This plasma is still hot by our everyday standards, but is also usually (but not always) cooler than the dense plasma below.

Those circumstances, with a cooler thin plasma above a hot, dense one, are perfect for the absorption phenomenon described previously. Photons of all energies bombard the atoms and ions in the thin plasma atmosphere of the star. Those with energies that correspond to transitions in the atoms or ions are absorbed, leaving atoms or ions that are 'excited' beyond the ground state; most of the rest escape the star to reach our eyes, telescopes, and recording devices.

The excited atoms or ions don't remain that way; their electrons fall back to the ground state, re-emitting photons of the energy that was previously absorbed. That would suggest that the absorption/re-emission process has no net effect on what we see, except for one aspect: the re-emitted photons travel in a random direction. As often as not, that's back toward the center of the star. The hurly-burly of the denser plasma includes processes that redistribute the energy of the single photon differently between photons, electrons, and ions. Most of the time, that redistribution leads to a photon of a different energy, which can then escape the star.

The end result is that photons of energies corresponding to transitions in the atoms or ions are 'missing' from the spectrum of light that reaches Earth. There are some photons at those energies, but not as many as there would be without the absorption process. If a prism is used to sort the photons from the star by energy[14], there will be dark lines in the spectrum corresponding to the absorbed energies (figure 2.8).

This part of the theory was well-understood by the late 19th century. Spectral lines had been observed in stars, and in many cases identified as corresponding to

---

[14] That's what's happening when an ordinary dime store prism makes rainbows out of sunlight. The light from the Sun is being sorted, with photons of different energies being sent in different directions. We perceive sorted photons of different energies as being of different colors, and thus see rainbows, or, using the Latin word for rainbows, 'spectra'.

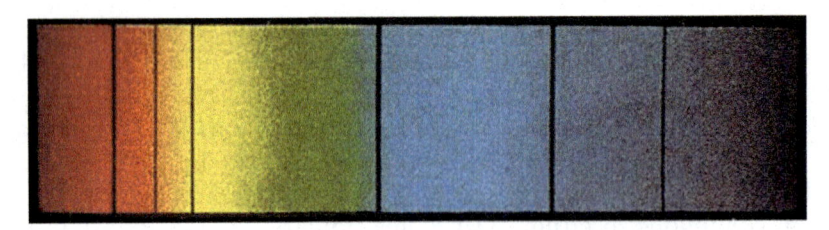

Figure 2.8. Example of a stellar spectrum showing absorption lines.

known elements measured in terrestrial laboratories. These identifications, however, led to new questions and theoretical developments.

For one thing, spectral lines do *not* occur precisely at a single energy. They have some 'width'—a narrow range of energies over which absorption occurs. The lines fade at either end, suggesting that absorption becomes less strong further away from the central energy for the line.

This is due primarily to three effects:
1. Lines have an inherent width. This is a manifestation of the Heisenberg uncertainty principle, which requires that an excited state not have a precisely defined energy.
2. If there are frequent collisions between the particles (atoms, ions, electrons) within the plasma, this shortens the lifetime of the excited states. This increases the uncertainty in energy under the Heisenberg uncertainty principle, and thus increases the width of the line.
3. If an atom or ion is moving toward the observer, it will absorb at a slightly higher energy, and if it is moving away, at a slightly lower energy. This is called the 'Doppler shift'. If a plasma is at a high temperature, then all the atoms and ions are in rapid motion in random directions, leading to a broadening of the line.

Another complication is that, when light comes to us from a star, it's coming from a variety of depths. For example, if there is a type of atom that is absorbing strongly, then most of the photons we see at its absorption energy are from atoms near the top of the atmosphere re-emitting photons—photons from deep in the atmosphere are unlikely to make it out. At different depths a stellar atmosphere is likely to be at different temperatures and pressures, and might have a different composition.

Payne's work was the first serious attempt to work backward from stellar spectra to determine the relative abundance of the elements (whether in the form of atoms or ions) causing the spectral lines.

Naïvely, one might expect elements with a greater abundance to show stronger lines, but careful consideration will show that this is often not the case. Atoms or ions that are fairly abundant may feature lines that are 'saturated', that is, all of the photons at the line's energy are absorbed and re-emitted at least once on their way out of the star. Whether there is just barely enough of the responsible element to accomplish that or a hundred times what is needed, the line will be just as dark. It's like trying to determine how much rain has fallen by leaving a piece of paper outside and seeing what fraction is covered in drops—once there's enough rain to cover the whole paper, the amount of rain ceases to matter.

There's also a crucial factor implicit in the discussion in this section to this point. Astrophysicists care about, for example, how much iron is in a stellar atmosphere. But part of the iron is present as atoms, part as singly-ionized iron, part as doubly-ionized iron, and so on. Each ionization state yields a different set of spectral lines.

Fortunately, the fraction of an element present in each ionization state is dependent only on the temperature and the pressure. This part of the theory was worked out in the papers by Saha, Fowler, and Milne.

This means that while, for a given abundance and pressure, at some temperatures the center of a line might be saturated, at other temperatures it might not be. And at some temperatures there might be so little of the atom in the corresponding ionization state that the line wouldn't be visible at all (depending, of course, on the sensitivity of the measurement).

The same can be said for the 'wings' of a line, that is, the energies away from its center where some absorption still occurs. Since less absorption occurs in the wings than at the center, it can be easier to find a temperature at which a point in the wing is unsaturated than it would be for the center.

Payne, following suggestions by the earlier theorists, looked for the spectral class at which either the center of a line or a point in its wings first appears (or disappears, depending on your point of view!). Under those circumstances ('marginal appearance'), the line (or its wing) is being formed just above the dense layer responsible for the continuous spectrum. This removes the difficulty that otherwise different lines might be probing different depths of the star's atmosphere, each of which would have different temperatures, pressures, and possibly compositions. By studying marginal appearance, all elements are probed at the layer just above the one responsible for the continuous spectrum, putting them on the same footing.

To proceed further, Payne had to make some simplifying assumptions. She assumed that the only difference between the stars of different spectral types that she was studying, at the level just above that responsible for their continuous spectrum, was temperature. This was certainly questionable when it came to pressure, but she argued in her dissertation that the expected differences should have only a modest effect on the spectra of the stars she was studying.

Payne's more significant assumption was that of uniform composition. Although her dissertation makes an argument for uniform composition based on the fact that most stars could be placed in a single series of spectral classes (i.e., the differences in spectra seemed to depend only, or at least primarily, on temperature), that did not preclude the possibility that composition varied in a systematic way from class to class. Her assumption, however, would be tested by her analysis of the data. If the assumption were correct, then the difference from spectral class to spectral class for each element should depend on temperature in the way that Saha, Fowler, and Milne predicted.

Using data on marginal appearance as well as data on the spectral class for which the maximum of each spectral line occurred (meaning, typically, the maximum for the wings since the center of the line would be saturated), Payne confirmed that the data behaved as if the composition were the same for all stars. At that point, she could finally use the Saha, Fowler, and Milne theory to find the relative abundance of the elements in the stellar atmospheres.

It should be clear from this description that Payne's dissertation wasn't notable for a singular piece of insight or a shocking piece of experimental data. Instead, she faced a problem for which the broad outlines were clear, but which involved an interlocking set of tenuous assumptions. It took extraordinarily incisive and thorough work to understand how all the various factors worked together, and what a convincing result would even look like.

A flash of insight can be once in a lifetime; an experimental result can be a lucky break. The kind of tour de force embodied by Payne's dissertation, however, does not happen by accident.

References

[1] Haramundanis K (ed) 1996 *Cecilia Payne-Gaposchkin: an Autobiography and Other Recollections* 2nd edn (Cambridge: Cambridge University Press) 70–238
[2] https://www.nobelprize.org/nobel_prizes/facts/
[3] Somerville M 1834 *On the Connexion of the Physical Sciences* (London: John Murray)
[4] Whewell W 1834 On the connexion of the physical sciences. By Mrs Somerville *Quarterly Review* **51** 54–68
[5] Holmes R 2014 In Retrospect: on the connexion of the physical sciences *Nature* **514** 432–3
[6] Secord J A 2014 *Visions of Science: Books and Readers at the Dawn of the Victorian Age* (Oxford: Oxford University Press)
[7] In Memoriam: Mary Somerville 1874 *Evening Hours: A Family Magazine* **1** 104–14
[8] Curie E 1937 *Madame Curie* (Garden City, NY: Doubleday)
[9] Mme Curie, Noted French Scientist, Wins Nobel Prize for Discoveries 1912 *Pensacola Journal* (May 1, 1912) The same brief article appears in at least one other paper; it may be an uncredited wire service story
[10] Percher Miles W 1885 Women 'Nobly Planned', or How to Educate Our Girls (lecture, Donaldson, LA, May 27, 1885) as reported in *Donaldsonville Chief* (Donaldsonville, LA), June 6, 1885.
[11] Sponsel A 2002 Constructing a 'revolution in science': the campaign to promote a favourable reception for the 1919 solar eclipse experiments *Brit. J. Hist. Sci.* **35** 439–67
[12] Women at Cambridge, October 29, 1921 *The Spectator* (Cambridge, UK)
[13] Payne C H 1923 Proper motions of the stars in the neighborhood of M36 (NGC 1960) *Mon. Not. R. Astron. Soc.* **83** 334–7
[14] Murray J A H 1971 *Oxford English Dictionary* (Oxford: Clarendon)
[15] Douglas, Allie Vibert (1894–1988) Queens Encyclopedia accessed June 17, 2017, www.queensu.ca/encyclopedia/d/douglas-allie-vibert
[16] Sobel D 2016 *The Glass Universe: How the Ladies of the Harvard Observatory Took the Measure of the Stars* (New York: Viking)
[17] Fine T 2016 Harvard College Observatory in Images *Harvard-Smithsonian Center for Astrophysics High Energy Astrophysics Division* accessed December 18, 2016 https://hea-www.harvard.edu/~fine/Observatory/ajcannon.html

[18] Reed H L 1892 Women's work at the Harvard Observatory *New England Magazine* **6** 165–76
[19] Pickering E C 1898 The Observatory *Annual Reports of the President and Treasurer of Harvard College 1896-97* (Cambridge: Harvard) 248–59
[20] Gilbert W S and Sullivan A 1879 *HMS Pinafore; or, The Lass that Loved a Sailor* (San Francisco: Bacon & Co.)
[21] Fine T 2016 The Harvard University Pinafore *Harvard-Smithsonian Center for Astrophysics High Energy Astrophysics Division* accessed December 18, 2016 https://hea-www.harvard.edu/~fine/Observatory/pages/play.html
[22] Long J D 1900 Memorandum Respecting the Functions of the Proposed Board of Visitors to the Naval Observatory *Report of the Superintendent of the United States Naval Observatory for the Fiscal Year Ending June 30, 1900* (Washington: Government Printing Office)
[23] Letter, H Shapley to E Upton, January 23, 1930 as transcribed at Harvard-Smithsonian Center for Astrophysics High Energy Astrophysics Division, accessed June 17, 2017 http://hea-www.harvard.edu/~jcm/html/shapley.html
[24] Dolan G 2016 The Post of Computer *The Royal Observatory Greenwich* accessed December 21, 2016 http:www.royalobservatorygreenwich.org/articles.php?article=1000
[25] Mitchell M 1877 The Need of Women in Science *Papers Read at the Fourth Congress of Women, Held at St George's Hall, Philadephia October 4-6, 1876* (Washington: Todd Brothers) 9–11
[26] Wright H 1949 *Sweeper in the Sky: The Life of Maria Mitchell, First Woman Astronomer in America* (New York: Macmillan)
[27] Mack P E 1990 Straying from their orbits: women in astronomy in America *Women of Science: Righting the Record* ed G Kass-Simon and P Farnes (Bloomington & Indianapolis: Indiana University Press) 72–116
[28] Pickering E C 1882 *A Plan for Securing Observations of the Variable Stars* (Cambridge: John Wilson & Son)
[29] Pickering E C 1906 The aims of an astronomer *Science* **24** 65–71
[30] Pickering E C 1890 The Draper Catalogue of Stellar Spectra Photographed with the 8-Inch Bache Telescope as a Part of the Henry Draper Memorial *Ann. Astron. Obs. Harvard College* **27** (Cambridge: John Wilson & Son)
[31] The Draper Catalogue of Stellar Spectra 1891 *The Observatory* **14** 357–9
[32] Pickering E C and Fleming M 1897 Miscellaneous Investigations of the Henry Draper Memorial *Ann. Astron. Obs. Harvard College* **26** 193–P.XI.2
[33] A Distinguished Dundee Lady 1893 *The Flaming Sword* **5** 363–4 (Attributed to *Dundee News*)
[34] Johnson C and Higgins M 2008 (last modified) Antonia Maury *Vassar Encyclopedia* https://vcencyclopedia.vassar.edu/alumni/antonia-maury.html
[35] Hennessey L 1998 (last modified) Annie Jump Cannon (1863-1941) *Wellesley College* http://academics.wellesley.edu/Astronomy/Annie/education.html
[36] *Second Conference, Harvard Observatory, 1898* 1910 *Publ. Astron. Astrophys. Soc. Am.* **1** 43–72
[37] Richardson Donaghe H 1898 Photographic flashes from Harvard Observatory *Pop. Astron.* **6** 481–7
[38] Paton Stevens Fleming W 1900 *Journal of Williamina Paton Fleming* Harvard University Archives http://nrs.harvard.edu/urn-3:HUL.ARCH:666402

[39] McKeen Cattell J 1906 *American Men of Science: A Biographical Dictionary* 1st edn (New York: Science Press)
[40] Rossiter M W 1982 *Women Scientists in America Volume 1: Struggles and Strategies to 1940* (Baltimore and London: Johns Hopkins University Press)
[41] Cannon A J 1919 Astronomical Fellowships for Women *Harvard College Observatory Circular* **214** 1–2
[42] Pickering E C 1918 *Seventy-Second Annual Report of the Director of the Astronomical Observatory of Harvard College for the Year Ending September 30, 1917* (Cambridge: Harvard University)
[43] Interview of C Payne-Gaposchkin by Owen Gingerich on March 5, 1968, Niels Bohr Library & Archives, American Institute of Physics, College Park, MD USA www.aip.org/history-programs/niels-bohr-library/oral-histories/4620
[44] Gingerich O The Most Brilliant PhD Thesis Ever Written in Astronomy *Harvard Square Library* accessed December 24, 2016 http://www.harvardsquarelibrary.org/biographies/cecilia-payne-gaposchkin-3
[45] Saha M N 1921 On a physical theory of stellar spectra *Proc. R. Soc.* A **99** 135–53
[46] Fowler R H and Milne E A 1923 The intensities of absorption lines in stellar spectra, and the temperatures and pressures in the reversing layers of stars *Mon. Not. R. Astron. Soc.* **83** 403–24
[47] DeVorkin D H 2000 *Henry Norris Russell: Dean of American Astronomers* (Princeton: Princeton University Press)
[48] Payne C H 1924 On the absorption lines of silicon in stellar atmospheres *Harvard College Observatory Circular* **252** 1–12
[49] Menzel D H 1924 A Study of line intensities in stellar spectra *Harvard College Observatory Circular* **258** 1–20
[50] Kirchoff G 1862 *Researches on the Solar Spectrum and the Spectra of the Chemical Elements* translated by ed Henry E Roscoe (Cambridge: Macmillan)
[51] Rowland H A 1902 Report of progress in spectrum work *The Physical Papers of Henry Augustus Rowland* (Baltimore: Johns Hopkins) 512–24
[52] Norris Russell H 1914 The Solar Spectrum and the Earth's Crust *Science* **39** 791–4
[53] Eddington A S 1918 On the Conditions in the Interior of a Star *Astrophys. J.* **48** 205–13
[54] Rosseland S 1925 On the Distribution of Hydrogen in a Star *Mon. Not. R. Astron. Soc.* **85** 541–6
[55] Comptom K T and Russell H N 1924 A Possible Explanation of the Behaviour of the Hydrogen Lines in Giant Stars *Nature* **114** 86–7
[56] Payne C H 1924 On the spectra and temperatures of the B stars *Nature* **113** 783–4
Payne C H 1924 On ionization in the atmospheres of the hotter stars *Harvard College Observatory Circular* **256** 1–8
Payne C H 1924 On the spectra of class O stars *Harvard College Observatory Circular* **263** 1–6
[57] Payne C H 1924 A synopsis of the ionization potentials of the elements *Proc. Natl. Acad. Sci. USA* **10** 322–8
[58] Irwin A 1903 Radcliffe College *Ann. Rep. President and the Treasurer of Harvard College, 1901-1902* (Cambridge: Harvard University) 312–5
[59] Kidwell P A 1996 An historical introduction to 'The Dyer's Hand' *Cecilia Payne-Gaposchkin: an Autobiography and Other Recollections* 2nd edn ed Katherine Haramundanis (Cambridge: Cambridge University Press) 11–38

[60] Kelly S E and Rosner S A 2012 Winifred Edgerton Merrill: 'She opened the door' *Not. AMS* **59** 504–12

[61] Bracher K 2012 Anne S Young: professor and variable star observer extraordinaire *J. Am. Assoc. Variable Star Observers* **40** 24–30

[62] Thirty-Second Meeting of the American Astronomical Society 1924 *Pop. Astron.* **32** 451–61

[63] Letter, Henry Norris Russell to Cecilia H Payne, January 14, 1925, in The Most Brilliant Ph.D. Thesis Ever Written in Astronomy," by Owen Gingerich, *Harvard Square Library*, accessed December 24, 2016. http://www.harvardsquarelibrary.org/biographies/cecilia-payne-gaposchkin-3

[64] Payne C H 1925 *Stellar Atmospheres: A Contribution to the Observational Study of High Temperature in the Reversing Layers of Stars* Harvard Observatory Monographs (Cambridge: Harvard University Press)

[65] Druyan A, Sagan C and Soter S 2014 Sisters of the Sun *Cosmos: A Spacetime Odyssey* hosted by Neil deGrasse Tyson, first aired on Fox Broadcasting, April 27, 2014

[66] Payne C H 1926 The stuff stars are made of *The Universe of Stars: Radio Talks from the Harvard Observatory* ed Harlow Shapley and Cecilia H Payne (Cambridge: Harvard University Press)

[67] Elvey C T and Struve O 1930 A study of stellar hydrogen lines and their relation to the Stark effect *Astrophys. J.* **72** 277–300

[68] Interview of H Hogg by Owen Gingerich on October 25, 1987, Niels Bohr Library & Archives, American Institute of Physics, College Park, MD USA. www.aip.org/history-programs/niels-bohr-library/oral-histories/29928-2

[69] Letter, E Hubble to H Shapley, February 19, 1924, in *Edwin Hubble: The Discoverer of the Big Bang Universe* by Alexander S Sharov and Igor D Novikov, 1993, translated by Vitaly Kisin (Cambridge: Cambridge University Press)

[70] Shapley H 1913 The orbits of eighty-seven eclipsing binaries—a summary *Astrophys. J.* **38** 158–74

[71] H Norris Russell to H Shapley, August 11, 1925, in The Most Brilliant PhD Thesis Ever Written in Astronomy by Owen Gingerich, *Harvard Square Library*, accessed December 24, 2016. http://www.harvardsquarelibrary.org/biographies/cecilia-payne-gaposchkin-3

[72] E A M 1925 Stellar atmospheres: a contribution to the observational study of high temperature in the reversing layers of stars *Nature* **116** 530–2

[73] Merrill P W 1926 Stellar atmospheres: a contribution to the observational study of high temperature in the reversing layers of stars, Cecilia H Payne *Astron. Soc. Pacific* **38** 33–4

[74] Stewart J Q 1925 Stellar atmospheres: a contribution to the observational study of high temperature in the reversing layers of stars, Cecilia H Payne *Phys. Rev.* **26** 900

[75] Struve O 1926 Stellar atmospheres. a contribution to the observational study of high temperature in the reversing layers of stars by Cecilia H Payne *Astrophys. J* **64** 204–8

[76] Struve O and Zebergs V 1962 *Astronomy of the 20th Century* (New York: Macmillan)

[77] Eddington A S 1926 *The Internal Constitution of the Stars* (Cambridge: Cambridge University Press)

[78] McKeen Cattell J and Cattell J 1927 *American Men of Science: A Biographical Dictionary* 4th edn (New York: Science Press)

[79] Women *Princetonia* last updated January 4, 2016

[80] Sawyer Hogg H 1984 Cecilia Payne-Gaposchkin: an autobiography and other recollections *Physics Today* **37** 67

[81] Astronomy Alumni *Department of Astronomy* accessed June 17, 2017, https://astronomy.fas.harvard.edu/astronomy-alumni
[82] A Brief History of the Harvard College Observatory *Department of Astronomy* accessed June 17, 2017, https://astronomy.fas.harvard.edu/astronomy-alumni
[83] Resurrected from Files of Long Ago *Cambridge Chronicle* January 17, 1930, as transcribed at *Harvard-Smithsonian Center for Astrophysics High Energy Astrophysics Division*, accessed December 26, 2016 http://hea-www.harvard.edu/~jcm/html/chron.html
[84] Haramundanis K 1996 A personal recollection *Cecilia Payne-Gaposchkin: an Autobiography and Other Recollections* 2nd edn ed Katherine Haramundanis (Cambridge: Cambridge University Press) 39–69
[85] Meldrum Stewart R 1930 The Forty-Third Meeting of the American Astronomical Society *J. R. Astron. Soc. Canada* **24** 49–54
[86] Forty-Third Meeting 1931 *Pub. Am. Astron. Soc.* **6** 331–6
[87] Conan Doyle A 1890 *The Sign of Four* (London: Spencer Blackett)
[88] Norris Russell H 1929 On the composition of the sun's atmosphere *Astrophys. J.* **70** 11–82
[89] Eddington A S 1932 The hydrogen content of stars *Mon. Not. R. Astron. Soc.* **92** 471–81
[90] Payne C H 1930 The Stars of High Luminosity *Harvard Observatory Monographs* (New York and London: McGraw-Hill)
[91] Payne-Gaposchkin C 1943 So you want to do research *Pro Tem* **1** 1–2
[92] Award of gold medals to Dr Annie J Cannon and Professor Henry B Bigelow 1930 *Science* **74** 644–7
[93] Carey C W 2006 *Am. Sci.* (New York: Infobase)
[94] McLaughlin D B 1935 The Fifty-Third Meeting of the American Astronomical Society *Pop. Astron.* **63** 71–8
[95] Aitken R G, Wyse A B and Gaposchkin S 1938 General Notes and reviews *Pub. Astron. Soc. Pacific* **50** 134–42
[96] Larrabee E 1950 The day the sun stood still *Harper's* January
[97] Payne-Gaposchkin C 1950 Nonsense, Dr Velikovsky! *The Reporter* 14 March 37–42
[98] Payne-Gaposchkin C 1950 Worlds in collision *Pop. Astron* **58** 278–86
[99] Menzel D H 1950 Worlds in collision *Phys. Today* **3** 26
[100] About the Collection *Astronomical Photographic Plate Collection* accessed December 26, 2016, http://platestacks.cfa.harvard.edu/about-collection

IOP Concise Physics

Beyond Curie
Four women in physics and their remarkable discoveries, 1903 to 1963
Scott Calvin

Chapter 3

Lise Meitner

3.1 Making up for lost time

Lise Meitner was born in Vienna in 1878. She would likely have had an easier life had she been born just a few years later.

In the nineteenth century, higher education for women was not allowed in central and eastern Europe. Universities were government entities, and thus subject to regulations governing who could enter.

Matters were different in the United States. From the founding of Wesleyan in 1836 and Mount Holyoke in 1837, women's colleges provided a route to the bachelor's degree. Even earlier, Oberlin pioneered coeducation from its founding in 1833. The example set by these institutions and others like them gradually eroded the barriers to women at top research universities. Maria Mitchell obtain a position for women to sit in on lectures at Harvard as early as 1868 [1], Winifred Edgerton received a special dispensation to receive a doctorate from Columbia in 1886 [2], and 'coordinate' colleges such as Barnard and Radcliffe allowed women to attend classes with men at top institutions without actually granting them degrees from the men's institution. It took this process, however, more than a century to break down all the formal barriers to higher education for women, with many male-only schools not becoming fully coeducational until the 1960s. The ability of women to become faculty at these universities took a similar trajectory.

Western Europe followed a similar, if somewhat delayed, pattern. The Sorbonne in France began admitting women in 1860, followed by Zurich University in 1867 [3]. In 1869, Girton College opened at Cambridge, joined six years later by Newnham (Cecilia Payne's eventual alma mater). By the time Meitner was a year old, Oxford had two women's colleges of its own: Lady Margaret and Somerville, the latter named after the same Mary Somerville discussed in the chapter on Cecilia Payne.

But in the Austro–Hungarian, German, and Russian empires, there were virtually no options [4, 5][1]. It was this that, in 1891, drove a 23-year-old Polish governess named Marya Sklodovksi to scrape together enough savings to leave her native Poland for the Sorbonne in Paris, where, after she Gallicized her first name and gained a new surname by marriage, she became Marie Curie [6].

It is possible that Meitner could have followed a similar path, eventually, although as an Austrian the more usual destination would have been Zurich, rather than the Sorbonne. In 1892, at the age of thirteen, she completed the public schooling available to Viennese girls. From there, she attended a private girls' high school designed to prepare her for teaching French. Like Curie, she could have worked as a teacher, tutor, or governess, striving to save enough to continue her studies abroad.

But it turned out she would not need to. In Austria, women were uniformly barred from higher education by an official regulation. Unlike in the United States or Britain, where resistance had to be overcome one school at a time, the role of government in higher education meant that in central Europe change would come by leaps and bounds: universities were opened to women in Hungary in 1895, in Austria in 1897, in the German state of Baden in 1900, Prussia in 1908, and the remainder of Germany in 1909 [7]. Meitner, it seemed, was completing high school at just the right moment.

Except that it was the wrong kind of high school. To enter a university she needed a college preparatory education, as demonstrated by passing the *matura*, a college-entrance exam. Her girls' high school had taught her little of what she needed to know.

And so, for two years after graduating high school, Meitner focused on making up for lost time, studying all the subjects she would need for the *matura*. In 1901, she passed the exam, allowing her to enter the University of Vienna that fall, at the age of 23.

Meitner soon chose experimental physics as her subject, completed her undergraduate degree, and then, in 1906, her doctorate, with no more than the usual amount of difficulty encountered by students of the subject. She was not the first woman to do so at the University of Vienna; Olga Steindler, who had the benefit of a high school education better suited to the *matura*, achieved that milestone three years before [8]. Meitner and Steindler worked in the same laboratory under Franz Exner.

After graduation, Meitner wrote to Curie, asking for a position in her laboratory, but received no response [5]. Instead, she remained in Vienna, performing experiments on alpha radiation under the guidance of Stefan Meyer, a young professor[2] at

---

[1] The Bestuzhev Courses in St Petersburg did provide something like a college education for women beginning in 1878. But university diplomas in the Russian Empire were state-sanctioned, so degrees from the Bestuzhev Courses were not recognized until 1910. There were also women who managed to secure degrees, even doctorates, by receiving a series of special exemptions.

[2] Specifically, a *Privatdozent*. The academic ranks of early 20th century Austria and Germany do not correspond neatly to modern academic ranks.

the University. This allowed her to publish and begin to develop a scientific reputation. Just a few years later, a donation from a wealthy industrialist allowed Meyer and Exner to establish the Institute for Radium Research, which soon became one of the world's leading institutions of nuclear physics. In addition, it would soon support an unusually large community of female scientists [9]. If Meitner had remained in Vienna, I have little doubt she would have been among them[3].

But Meitner was restless. To her, Vienna appeared to be a scientific backwater: its Institute for Theoretical Physics was 'located in a very primitive, converted apartment house' with an 'entrance that looked like a hen house', the introductory laboratory course, likewise, included only 'extremely primitive apparatus' and was taught by an instructor who was 'rather skeptical of the modern developments of physics' [10]. In addition, the University of Vienna had been home to one truly famous physicist, the mercurial Ludwig Boltzmann. Boltzmann was an inspiration and role-model for Meitner, and during her student days she grew close to him and his family. In late summer of 1906, Boltzmann hanged himself. It is little wonder that Meitner did not see her future lying in Vienna.

But where to go? Boltzmann had spoken enviously of German laboratories, and told his students that he regretted not accepting the offer of a professorship in Berlin [11].

After Boltzmann's death, Professor Max Planck came to Vienna from Germany to consider becoming his successor, but decided to decline the offer and remain at Berlin [12].

Beyond these influences, Meitner did not research her options or analyze her choice. She did not even realize that the state of Prussia, and therefore the Friedrich Wilhelm University in Berlin, did not yet admit women [10]. On little more than a hunch and the financial support of her parents, Meitner left for Berlin in 1907.

Once in Berlin, she convinced a somewhat bemused Planck to allow her to sit in on his lectures and then asked Professor Heinrich Rubens if she could have a space to do research in his lab. Both requests were granted.

It was also through Rubens that Meitner was first led to collaborate with Otto Hahn [13].

Hahn was a nuclear chemist, just a few months younger than Meitner. Like her, he was still supported financially by his parents. He had an easy-going, self-deprecating charm. Having just achieved the rank of *Privatdozent* a few months before [14], he outranked Meitner, but since she was a physicist and he a chemist and thus brought different skills and perspectives to their work, they considered themselves scientific equals from the start.

---

[3] Ruth Sime, in her biography of Meitner, suggests that, had Meitner remained in Vienna, she would likely never have risen to the kind of leadership position she eventually found in Berlin. In my opinion, this may be underestimating the effect of Meitner's particular talents—it seems equally unlikely that, had she not gone to Berlin, another woman there would have risen as she did. In the alternate reality where Meitner stayed in Vienna, a historian might easily speculate that, had she gone to Berlin, it was unlikely that she would have risen to a position of leadership!

It would only take a few years for Meitner's official status to catch up to Hahn's. By 1909 the doors of Friedrich Wilhelm were officially opened to women, allowing Meitner more freedom to use the facilities. In 1912, she became Planck's paid teaching assistant[4], the first woman in Prussia to achieve that rank [11]. 1912 also marked the opening of the Kaiser Wilhelm Institute for Chemistry (KWI for Chemistry), a major research institute financed and administered as a public–private partnership. Hahn was appointed as a scientific associate at its opening; a year later, Meitner was given the same title. Their research group was known as the Hahn–Meitner Laboratory, an equal partnership between a chemist and a physicist.

At first, however, their salaries were not comparable, with Hahn earning around 4000 marks per year and Meitner only 1500[5]. When another university courted her, Meitner's salary was doubled to 3000 marks per year.

In 1917, Meitner was named to head the new physics section within the KWI for Chemistry, and her salary was raised again, this time to match Hahn's. In 1919, she was given the title Professor—a special honor in imperial Germany, roughly equivalent to 'distinguished professor' in current American usage.

Despite the lost time caused by the delays in admitting women to higher education in Austria and Germany, by the age of forty Lise Meitner had caught up. But she did not stop there. In 1920, women were finally allowed to join the faculty in the Prussian public universities as well, soon providing her an appointment at the Friedrich Wilhelm University which allowed her to teach and supervise students, likely a first for a woman in the German Empire [5]. Her research group became one of the leading centers of nuclear physics in the world [15], attracting scientists from as far away as China [16]. When Payne was writing her doctoral dissertation at Harvard, Meitner was already an internationally respected scientist with influence and power (figure 3.1).

3.2 Questions of credit

In the field of nuclear physics, the question of who should get credit for a 'discovery' has been fraught almost from the beginnings of the field. Discoveries of new phenomena often involved what the philosopher of science Thomas Kuhn has called a 'paradigm shift'; a change in established theory needed in order to interpret the new phenomenon. Kuhn describes how this could complicate the attribution of discovery. One of his examples was a result from a previous century, the discovery of oxygen:

> Priestley's claim to the discovery of oxygen is based upon his priority in isolating a gas that was later recognized as a distinct species. But Priestley's sample was not pure, and, if holding impure oxygen in one's hands is to discover it, that had been done by everyone who had bottled atmospheric air.

---

[4] In the sense that she graded papers; the German rank of *Assistent* did not include the right to instruct students.

[5] Hahn also earned a 'marriage supplement' of about 1000 marks per year. Had Meitner been married, it is unclear whether she would also have been entitled to that additional income.

Figure 3.1. Meitner's scientific colleagues in 1920. Meitner is the woman seated furthest to our right on the sofa; Hahn is at our far right in an easy chair.

Besides, if Priestly was the discoverer, when was the discovery made? In 1774 he thought he had obtained nitrous oxide, a species he already knew; in 1775 he saw the gas as dephlogistated air, which is still not oxygen or even, for phlogistic chemists, a quite unexpected sort of gas…If we refuse the work to Priestly, we cannot award it to Lavoisier for the work of 1775 which led him to identify the gas as the 'air itself entire.' Presumably we wait for the work of 1776 and 1777 which led Lavoisier to see not merely the gas but what the gas was. Yet even this award could be questioned, for in 1777 and to the end of his live Lavoisier insisted that oxygen was an atomic 'principle of acidity' and that oxygen gas was formed only when that 'principle' united with caloric, the matter of heat. Shall we therefore say that oxygen had not yet been discovered in 1777? Some may be tempted to do so. But the principle of acidity was not banished from chemistry until after 1810, and caloric lingered until the 1860s. Oxygen had become a standard chemical substance before either of those dates [17].

Kuhn's discussion illustrates the difficulty in giving attribution to discoveries that involve new observations which require modification of existing theory, but complications exist even when new theory is not needed. For an example from a different branch of science, consider the question of the discovery of a 'new' biological species. In general, these species are not new in that they are newly arisen, only in that they are newly identified. Presumably members of the species had been seen many times before. It remains necessary for a scientist to assert that the species exists, and then to provide evidence. But what evidence is sufficient: a single dead organism? A fossil fragment? A photograph? A drawing? The answer to each of those questions is 'yes, sometimes'. Any requirement which is too strict would leave out some examples that are clearly representatives of newly discovered species, but if we are insufficiently strict then we will end up with many more different 'species' than actually exist in nature.

In cases such as speciation or, to take an example from nuclear physics, the discovery of a 'new' element, the scientific community has to make a collective decision (perhaps in the form of the vote of an international committee). That decision is necessarily to some extent subjective.

Subjectivity, in turn, opens the door to a multitude of influences: politics, personal animus, explicit and implicit bias.

By making that observation, I don't mean to impugn the value of science itself, to imply that it doesn't reflect actual properties of the world around us, nor do I mean to suggest that science does not, over time, provide an increasingly accurate and thorough description of our universe. But deciding who deserves the 'credit' for each discovery is a different kind of question. In a rigorous sense, it is usually unanswerable, for as Kuhn points out discovery is a process, not an instant.

There is a particularly subtle and pernicious way in which bias can effect the attribution of credit. Often, credit for a new theory is given to the scientist who presents an argument that is convincing to the rest of the scientific community. Newton, for example, provided convincing arguments that gravity makes the planets orbit the sun, although he was not the first to make the suggestion. Likewise, Russell was the one who finally convinced astronomers that the atmosphere of the sun is predominately hydrogen.

Suppose credit of that kind were bestowed by a vote: if 90% of the members of a scientific society felt convinced, then the person would be credited with the discovery. This kind of system is not entirely hypothetical; it has similarities, for instance, to the method by which Nobel Prizes are awarded (described in detail later in this chapter).

Now further suppose that 15% of the voting members harbor substantial bias toward a particular group: women, say, or Jews, or blacks. Even if the other 85% held no biases whatsoever, the members of the group being discriminated against would face an almost insurmountable obstacle toward gaining recognition.

3.3 A scientific powerhouse

Regardless of what particular discoveries Meitner is 'credited' with, it is clear that in the prime of her career she was one of the leading nuclear physicists of the time, involved in the discovery and verification of many new phenomena and explaining others. An exhaustive list of her scientific accomplishments is beyond the scope of this current work, but highlights include:
- Demonstrating that alpha particles are deflected when they interact with matter, a predecessor to Rutherford's famous gold foil experiment [18].
- The development, with Hahn, of the 'recoil method' for separating the products of radioactive decays from the mother substance [19].
- The discovery [20, 21], with Hahn, of ^{207}Tl, ^{208}Tl, ^{210}Tl, ^{233}Th, ^{231}Pa, ^{233}Pa, and ^{239}U.

- Preliminary evidence [22], with Hahn, for ^{214}Po.
- The naming, with Hahn, of the element protactinium [14].
- Being among the first to speculate on the existence of nuclear energy levels [11].
- Demonstrating that gamma emission followed beta decay rather than being triggered by it [12].
- Being the first to publish observational evidence for the Auger effect, in which the energy released when an electron falls into an unoccupied energy level of an ion is used to eject other electrons from the same atom [23].
- Noting that even atomic numbers tended to be more stable [24], foreshadowing Mayer's later work with magic numbers (see chapter 5).
- With Hahn, using the radioactive decay of uranium to lead to estimating the age of the Earth and thus the Sun, leading them to suggest that the Sun was powered by the conversion of mass to energy [11].
- Being among the first to identify pair production of positrons from a terrestrial source [25].
- Providing an improved determination of the mass of the neutron.
- Being among the first to realize the importance of neutron energy to the artificial transmutation of elements [11].

Any three of these would have marked Meitner as a leading scientist. To have accomplished all of them, particularly considering the institutional barriers raised up against her gender early in her career, put her in the first rank of nuclear physicists of the era.

3.4 Tumult

Meitner's career spanned a particularly tumultuous periods in the history of Germany.

When she first came to Berlin, Germany was still an Empire under the bellicose leadership of Kaiser Wilhelm II. She had only been in Berlin a few years when World War I broke out. As a citizen of Austria and a resident of German Prussia, Meitner at first viewed the war with optimism, even enthusiasm [11]. In enemy France, Marie Curie immediately set about forming a medical x-ray service for military use, soon consisting of 200 stations and 20 mobile units [6]. At the start of the war, in 1914, Meitner was still too junior to take that kind of organizing role, but she did volunteer to serve in Austrian x-ray units near the front lines beginning in 1915. Serving at the Russian front, then the Italian, and then the Russian again, she also spent time during the war continuing her research at the KWI for Chemistry, and, for one period, back at Stefan Meyer's Radium Institute in Vienna [11].

In 1918, the other Central Powers of Bulgaria, the Ottoman Empire, Austria, and Hungary surrendered one by one, leaving Germany alone and facing certain defeat. In the last days of the war, a naval mutiny rapidly became a revolution.

For months, Germany was in chaos. The Kaiser had abdicated, allowing a brutal and violent power struggle between communist, nationalist, and various centrist groups. The centrist coalition eventually prevailed, and the Weimar Republic came into being. The fledgling republic was forced by the victorious allied powers to agree to the Treaty of Versailles, a humiliating and debilitating blow. Instability would continue for years, including coups, assassinations, and revolts.

Meanwhile, the crushing terms dictated by the Treaty of Versailles, along with the instability of the German government, led to severe hyperinflation. Over the course of a few years, the value of the German mark fell by a factor of a *trillion*.

Finally, in the mid-20s, some stability returned to both the government and the economy.

Throughout empire, war, revolution, and economic crisis, the KWI kept fulfilling its mission of scientific research. From the end of the war onward, many scientists around the world saw internationalism as the answer to the catastrophe of world war, and strongly supported their colleagues in the defeated nations.

Year after year new crises would arise, and the scientists at the KWI would find new ways of coping. 'This too shall pass' could have been their motto. It is little wonder that, when Adolph Hitler of the Nazi Party became Chancellor in January of 1933, Meitner and her colleagues hoped that this was just another crisis that they could wait out [26]. Meitner's nephew Otto Frisch, himself a young physicist at the University of Hamburg, has described his own attitude at the time:

In the early thirties in Hamburg I didn't pay any attention to the general crisis atmosphere; with a sarcastic smile I observed the repeated changes of government and the much joked-about ineptness of Hindenburg, the famous general who had been made President of the Republic of Germany. When a fellow by name of Adolph Hitler was making speeches and starting a Party I paid no attention. Even when he become [sic] elected Chancellor I merely shrugged my shoulders and thought, nothing gets eaten as hot as it is cooked, and he won't be any worse than his predecessors [27].

For Frisch, that attitude did not last long. By the end of the year, he had been forced out of his job by the Nazis and had taken up a new position with Niels Bohr in Copenhagen.

Meitner did not feel herself to be in any immediate danger, although she was of Jewish ancestry. She was protected by multiple aspects of her status:
- She had converted to Protestantism in 1908 [11] and was not thought of as Jewish by her colleagues [28].
- She was an Austrian citizen, with an Austrian passport.
- She was a Great War veteran who had served one of the Central Powers.
- She had begun work at the KWI for Chemistry prior to the Great War.
- The KWI for Chemistry was a private–public partnership funded largely by private companies, as opposed to a government institution like the University.
- She had not been politically active.
- She had no enemies, and had important supporters. In particular, since 1928, Hahn had been head of the KWI for Chemistry. In addition, Max Planck was

the president of the Kaiser Wilhelm Society, the KWI for Chemistry's parent organization. Beyond that, Meitner's friends included a who's-who of leading physicists: Albert Einstein, Bohr, Erwin Schrödinger.
- She had an international reputation as an important scientist.

In addition, Hitler initially became chancellor through the democratic process. While the Nazi Party had not received an outright majority of votes, with under 40% of the vote in each of the 1932 elections, it was the largest single vote-getter in Germany's multi-party parliamentary system. After the appointment of Hitler as chancellor he called new elections, but despite intimidation and violence by his followers and arrests by his government, the Nazi Party still received under 50% of the vote.

It was, however, enough. The Nazis formed a coalition government, and then proceeded to arrest members of the communist party. Hitler called a new election, but even with the communists out of the way and intimidation by his followers, the Nazi Party still could not achieve 50% support.

Finally, Hitler proposed a law which would, in effect, suspend the constitution and give him near absolute control of the government. He detained many opposing members of parliament and negotiated with others, promising safeguards he would not, in the end, deliver. The 'enabling act' passed. By the fall of 1934 the remaining vestiges of the Republic had been swept away. Germany had become a dictatorship.

While some scientists with weaker ties to Germany, including Einstein, Schrödinger, and Meitner's nephew Frisch, left during this period, many stayed [29]. Meitner would have been welcomed in many laboratories across the world, but she would not have had her own section, painstakingly developed by her own effort, like she did in Berlin. After more than twenty years in Germany, after outlasting an empire, a war, a revolution, and a republic, she had no intention of leaving.

One by one, however, her protections were stripped away.

First, in April of 1933, the 'Law for the Restoration of the Professional Civil Service' [30] made clear that Jewish heritage was enough to make Meitner 'non-Aryan'. Her baptism as a Protestant was irrelevant. The law did, however, allow an exception for those 'who had fought in the World War at the front for the German Reich or for its allies'. Whether that clause included Meitner was at first open to interpretation.

In May, it became clear that her service in World War I was, in the eyes of the Nazis, insufficient. The Third Ordinance on the Implementation of the Law for the Restoration of the Professional Civil Service included the following clarification:

Front-line fighters in the meaning of the law include anyone who has participated in the World War (in the period from August 1, 1914 until December 31, 1918) as a combat soldier in battle, or who has taken part in a skirmish, in trench fighting, or in an occupation force…it is not sufficient for someone to have stayed in the war zone during the war for official reasons without having confronted the enemy [31].

This ordinance essentially prevented women who had served in capacities such as nurses or x-ray technicians from receiving the exemption. Another protection was gone.

By September, it was official: Meitner had lost her position at the University. Her service in the Great War was not, according to the government ruling, 'at the front', her position as Planck's *Assistent* insufficient to qualify. She retained, however, her primary position at the KWI for Chemistry, and the protection of Hahn and Planck.

Meanwhile, science itself was under attack. Since the early 1920s, the Nobel Laureate physicists Philipp Lenard and Johannes Stark had been advocating the superiority of 'Aryan physics', rejecting ideas such as relativity and quantum mechanics as 'Jewish fraud' [32]. It is no slur to refer to the pair as Nazis. In 1924, while Hitler, in jail for treason, dictated *Mein Kampf* to his co-conspirators Rudolph Hess and Emil Maurice, Lenard and Stark published a letter pledging their support:

> We need lucid minds as scientists—no, people should not be segregated according to their occupation (a frequently used Semitic deception!). We want lucid, well-rounded characters, just as Hitler is one. He and his comrades in the struggle appear to us as God's gifts from times of old when races were purer, people were greater, and minds were less deluded. This we feel; and these divine gifts should not be taken from us. This thought alone should be a solid enough basis to hold the nationally-minded together toward their great goal: Founding a new Germany, with Hitler 'beating the drum', in which the German spirit is not just tolerated again to a certain extent and released from imprisonment, no, but in which the German spirit is protected, nursed, and assisted so that it can then finally thrive again and develop itself further for the vindication of the honor of life on our planet which is now dominated by an inferior spirit [i.e., a Jewish one]. Universities and their students have failed, most of all precisely in those subjects that should have set the pace long ago. But it is also much better that 'the man of the people' is doing it. He is here. He has revealed himself as the '*Führer*' of the sincere. We shall follow him [33].

Once Hitler came to power, Lenard and Stark attempted to put their ideas in to action, scrubbing 'Semitic' science and scientists from the record. Those who pushed back were labeled 'white Jews':

> When the carrier of this mentality is not a Jew but a German, it is twice as important to fight him than it is to fight an ethnic Jew, who cannot conceal the source of his mentality [34].

In the face of all of this, what was the 'right' course of action for scientists like Planck and Hahn and Werner Heisenberg, all of whom loved Germany and opposed the Nazis? Protect the vulnerable, or the important, or those closest to them, or the science itself? Resign their posts in protest, only to see a Nazi intent on the destruction of science and scientists replace them, or remain in an attempt to salvage what they could which would have the side effect of lending legitimacy to the regime and its philosophy? Every tactic, from literal suicide to obstructionism to the attempt to win the trust of Hitler, was tried by someone. These attempts were often successful in their immediate goal, protecting a person here or a bit of scientific integrity there [35]. But it was like trying to hold back the tide.

Planck, in particular, fought against the Nazification of physics and the elimination of all the remaining 'non-Aryans' from the KWIs. Lenard and Stark had hoped he would retire from the presidency of the Kaiser Wilhelm Society in 1933, but he remained. In 1936, it looked like either Stark or Lenard would succeed Planck as president, but Planck stayed on for one more year [32]. When he finally stepped down, Planck arranged for his replacement to be Carl Bosch, a Nobel laureate and powerful industrialist who had opposed Nazi meddling with science [36]. But the price was that Bosch's new deputy was to be the Nazi Ernst Telschow. Telschow had been a doctoral student of Hahn's [14], but would that loyalty extend to allowing the continued employment of a non-Aryan?

Meitner's defenses were holding, but barely. By the end of 1937, the list now looked like this:

- ~~She was a Protestant.~~
- She was an Austrian citizen, with an Austrian passport.
- ~~She was a Great War veteran who had served one of the Central Powers.~~
- ~~She had begun work at the KWI for Chemistry prior to the Great War.~~
- The KWI for Chemistry was a private–public partnership funded largely by private companies, as opposed to a government institution like the University.
- She had not been politically active.
- She had no enemies, and had important supporters. In particular, since 1928, Hahn had been head of the KWI for Chemistry. ~~In addition, Max Planck was the president of the Kaiser Wilhelm Society, the KWI for Chemistry's parent organization. Beyond that, Meitner's friends included a who's-who of leading physicists: Einstein, Bohr, Schrödinger.~~ Most of her prominent supporters had either left Germany or retired, and some, such as Einstein, were despised by the Nazis.
- ~~She had an international reputation as an important scientist.~~ Her international reputation as a scientist was intact, but it was becoming increasingly less clear that this would be seen as an asset by the Nazi regime.

In March of 1938, Germany annexed Austria, removing Meitner's last formal defense. Within a day Nazis within the KWI for Chemistry were calling for her removal [11].

For a few weeks, her relationships held, but it was clear she was in great danger of losing her job. In retrospect, we now realize she was in danger of losing much more than that, but at the time she thought she could still emigrate freely; early in the Nazi era, Jews had been encouraged to leave.

But her Austrian passport presented a problem. She soon decided to leave for Bohr's institute in Copenhagen, but found the Danish government no longer accepted Austrian passports as valid, and the German regime's attitude toward emigration was changing. Since late 1937, the regime had quietly been making it more difficult for Jews to obtain passports [37]. In Meitner's case, a decision came down in June:

> It is considered undesirable that well-known Jews leave Germany to travel abroad where they appear to be representatives of German science, or with their names and their corresponding experience might even demonstrate their inner attitude against Germany [11].

Meitner was running out of options. And yet she was still working, still running the physics section at the KWI for Chemistry, even as she and her friends and colleagues around the world tried to find a way to get her out. Slowly, and almost too late, a plan came together.

On 12 July 1938, Meitner spent what to outward appearances was a normal work day at the KWI for Chemistry. Of the scientists working there, only herself and Hahn knew it was to be her last.

She and Hahn went to her apartment and packed a few things; she then spent the night at Hahn's house with his family, where he provided her a diamond ring he had inherited from his mother in case she needed something of value (for those who lived through the hyperinflation of the 20s, paper money did not seem entirely reliable). Neither one realized that their preparations had been seen and understood by Meitner's next-door neighbor Kurt Hess, a Nazi scientist who also worked at the KWI for Chemistry. Hess alerted the police, who fortunately did not act quickly. The next morning, Meitner took a train to the Netherlands. On the way, police repeatedly checked the papers of the passengers, arresting several of them. At one point, her now invalid Austrian passport attracted considerable attention. But the rapidly changing policies of the Nazi regime, masked by varying levels of secrecy and always subject to individual exceptions, ironically made arrest a crapshoot, rather than a certainty. While that short train trip held many tense moments for Meitner, she arrived in the Netherlands safely [15].

From there, Meitner made her way to Sweden, where she took a position in the Nobel Institute for Experimental Physics [12]. For Meitner, her new position was a significant demotion, and one she never fully adjusted to. Accustomed to running a section of a major research laboratory, overseeing dozens of assistants and graduate students and having access to machinists and an equipment budget of her own, in Sweden she was treated as a distinguished visiting scientist, not participating much in existing projects of the Institute but not provided the resources to effectively pursue her own. Theorists transplant with relative ease, at least scientifically; Einstein could do his work as well in Princeton as in Berlin. But established experimentalists, torn away from their laboratories and assistants, risk frustration and irrelevancy.

It is, of course, not impossible for leading experimentalists to make transitions of that sort. It is instructive to compare Meitner's transition with that of Enrico Fermi. In 1938, Fermi was a peer of Meitner, and of about the same scientific stature. Both led important research groups of comparable size, Meitner the Physics section of the

KWI for Chemistry, and Fermi the Physics Institute at the Sapienza University of Rome. As part of Mussolini's alliance with Hitler, the rights of Jews were being curtailed in Italy. Fermi's wife, Laura, was Jewish, and so they decided to emigrate to the United States [38]. Almost immediately upon coming to the New York, Fermi began work at Columbia, and was soon leading a team of nuclear physicists, with access to generous opportunities for collaboration, along with ample support in the form of assistants, students, equipment, and funding.

What accounts for the difference? While similar in many ways, Fermi had three significant advantages over Meitner:
1. He was more than twenty years younger, and thus could be expected to have a longer productive career in front of him.
2. He was male.
3. On 10 November, 1938, he was awarded the Nobel Prize in physics [39].

3.5 How Nobel Prizes are selected

The Nobel Prize is widely considered the most prestigious award in both physics and chemistry. It is therefore worth examining the selection process for the physics award; the process for chemistry is completely analogous.

The process began well before the establishment of the Prize, with the formation of the Royal Swedish Academy of Sciences in 1739 [40]. Since then, the Academy has been responsible for recruiting new members into its ranks.

At any given time, there are supposed to be 175 non-Swedish members of the Academy, and 175 Swedish members under the age of 65. Once a Swedish member turns 65, a new member is added, but the old member remains until death. Thus, there are generally many more Swedish members than non-Swedish members.

The Academy is further divided into sections by academic discipline. The sections, like the Academy as a whole, have fixed membership sizes. There are thus always supposed to be 18 members of the Academy who are non-Swedish physicists, and 18 Swedish physicists under the age of 65, in addition to Swedish physicists past that age [41]. At the time of this writing, there are 17 non-Swedish physicists (presumably with one vacancy) and 53 Swedish physicists who are members of this section. The non-Swedish members show disproportionate representation from countries bordering Sweden, consisting of one Dane, one Norwegian, three Finns, two Russians, one German, one Frenchwoman, one Scot, two Japanese, and five Americans. In addition, the Frenchwoman, the German, one of the Finns, and one of the Americans currently work in Sweden, while another of the Americans is of Scandinavian descent [42]. For the most part, this uneven geographic distribution (figure 3.2) has *not* led to an excessive concentration of prizes for Scandinavia and its neighbors [43]; the Academy has taken pains to make sure that the Prize is international in nature, as is specified in Nobel's will [44]. But it does mean that certain perspectives are more likely to be represented; while Swedish physicists are only modestly overrepresented among laureates, the *kind* of physics that Swedish physicists valued is likely to be more strongly favored [45]. For example, through the first half of the 20th century most of the prizes went for experiment rather than theory [43].

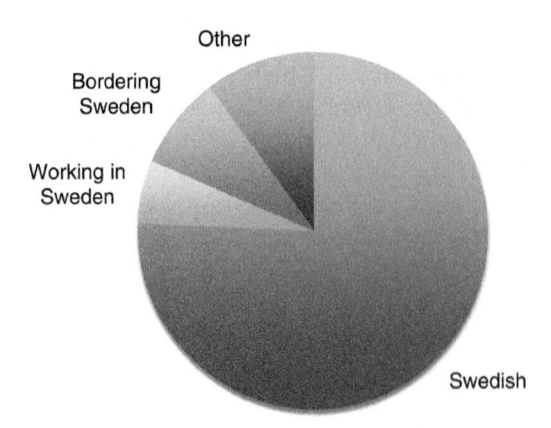

Figure 3.2. Composition of physicists within the Royal Swedish Academy of Sciences. 'bordering Sweden' includes maritime borders, but does not count members already counted as 'working in Sweden.'

The Academy then appoints the Nobel Committee for Physics. Nominally a five-member committee with each person holding a three-year term, additional 'adjunct' members are sometimes named, and the three-year terms are frequently renewed. While it is not required that the committee consist of physicists within the Academy, in practice that is generally the case. In 1938 the committee consisted of five male Swedish physicists, including Manne Siegbahn, the Director of the Nobel Institute of Experimental Physics at which Meitner now worked [43].

The Nobel Committee for Physics is responsible for soliciting nominations for the Prize from a wide range of international physicists. Qualified nominators always include members of the Academy, Nobel Laureates in Physics, and tenured professors from Sweden, Norway, Denmark, Finland, and Iceland. The Committee takes seriously its task of expanding this list and ensuring broad geographic representation, typically sending out thousands of requests for nominations.

Once they have received the nominations, the Committee performs a preliminary screening, commissioning reports, often by individual members of the Committee, on those they wish to move forward for consideration. The Committee then makes its recommendations, sending a report to the physicists of the Academy for discussion and a recommendation. Finally, the Academy as a whole votes.

The entire proceedings, from nomination to vote, are held confidential for fifty years [46]. After fifty years, the reports and nominations are released, but the deliberations surrounding the final vote by the full Academy remain confidential forever.

While the nomination process involves thousands of scientists around the world and the final vote involves hundreds (primarily but not exclusively from Sweden), the five-member Nobel Committee for Physics is a crucial gatekeeper in the process.

From 1924 to 1938, Meitner had been nominated a remarkable 19 times for a Nobel Prize (17 times for chemistry and twice for physics) [47], always jointly with

Hahn, but neither had yet received a prize. Since nominations are not made public for fifty years and even then are held confidential if the nominee is still alive, there is no contemporaneous cachet associated with being 'Nobel-nominated'.

3.6 Beyond uranium

As far as elements that occur naturally in quantities large enough to mine, uranium is the end of the periodic table. Everything we're familiar with, on Earth or in the heavens, is made up of some combination of the 92 elements from hydrogen (atomic number $Z = 1$) to uranium ($Z = 92$). Thus, when Irène Curie (daughter of Marie) and her husband Frédéric Joliot showed in early 1934 that previously unknown isotopes could be created by bombarding known isotopes with subatomic particles [48], the possibility of creating new elements arose [49][6].

Fermi's group, meanwhile, had decided to change their focus from spectroscopy to nuclear physics. Fermi's colleague Franco Rasetti spent much of the period from 1931 through 1934 in Meitner's laboratory, learning technical skills which he brought back to Rome [16].

After the Joliot–Curies[7] published their results, Fermi and his group began a systematic investigation of the transmutation of each element in the periodic table by bombarding them with neutrons, beginning with hydrogen and working their way up one by one, skipping only a few especially rare elements. Finally, they reached uranium. This yielded products which corresponded to none of the elements between radon ($Z = 86$) and uranium ($Z = 92$). Since all previous transmutations had resulted in a change in atomic number of no more than two at a time, this suggested that the newly produced isotopes must lie *above* uranium in the periodic table, at $Z = 93$ and 94. Fermi announced the likely production of the first transuranic elements, never before seen in nature.

This conclusion was nearly universally accepted. The *New York Times* proclaimed:

Fermi has succeeded in creating an artificial radioactive element which must be given the number 93 is of the highest importance... [sic] Fermi creates something so new that a niche has to be made for it in the table of elements. Transmutation, with its implication of changing a base to a noble metal, is transcended. To outdo nature and give us an element to be found nowhere on earth—alchemy seems less romantic and improbable in comparison [50].

Extraordinary claims require extraordinary evidence, however, as some scientists, among them the chemist Ida Noddack, pointed out.

Noddack was a chemist of international stature at the Physikalisch-Technische Reichsanstalt in Berlin (the same institute at which Frisch worked for a time), with

---

[6] Rutherford, much earlier, had demonstrated that bombardment could produce known, stable isotopes.

[7] In publication, both Irène and Frédéric used the surnames they had held prior to marriage, but in their private lives they both used the hyphenated form Joliot–Curie. I will therefore refer to them singularly by the names they published under, but jointly as the Joliot–Curies, a space-saving technique employed by many of their contemporaries.

the co-discovery of the element rhenium to her credit. By 1934 she had been nominated twice for the Nobel Prize, although in both cases the nominators indicated she was their second choice [51].

While Meitner knew Noddack professionally, it is clear that she did not count her among her friends and did not think highly of her standards, later referring to her 'unscientific small-mindedness [11].'

In response to Fermi's discovery, Noddack published a paper which, while careful not to criticize Fermi explicitly, critiqued the celebration in the popular press as premature:

> One must await further experiments, before one could claim that element 93 has really been found. Fermi himself is careful in this respect, as has been mentioned previously, but in [a news item in *Nature*] and also in the reports found in the newspapers it is made to appear that the results are already certain [52].

In her article, Noddack reasoned that, since nuclear physics had yielded surprises before, it might again. For instance, the product of the bombardment of uranium with neutrons might decay quickly through a long series of isotopes with very short half-lives. Even though each decay would change the atomic number by no more than two, within a matter of seconds the atomic number of the remaining product might be much lower than that of radon. To rule that out, Fermi would need to check for the presence of all elements, not just those from $Z = 86$ to $Z = 92$. Or something more unprecedented might have happened, such as the shattering of the uranium nucleus into large fragments.

Noddack's objections were considered briefly, but laid aside. It was the discovery of the transuranic elements that led to Fermi receiving a remarkable 34 Nobel nominations from 1935 to 1938, culminating in his 1938 prize for 'his demonstrations of the existence of new radioactive elements produced by neutron irradiation, and for his related discovery of nuclear reactions brought about by slow neutrons [53].'

Shortly after Fermi's results were published in 1934, Meitner asked Hahn to join a new collaboration with her, investigating the transuranes Fermi had discovered [54]. Their investigative team soon grew to include other scientists in their laboratories: in 1935, Fritz Strassmann, a chemist and principled opponent of the Nazis whom Hahn and Meitner had found a way to keep at the KWI for Chemistry (although desperately underpaid); and, in 1936, Clara Lieber, a doctoral student in chemistry from the United States by way of England [55].

Fermi's group knew that they lacked the expertise of Meitner and Hahn when it came to this kind of investigation:

> At this point Hahn and Meitner entered the game and confirmed our conclusions for what concerns the transuranic elements...after the confirmation of our results from Hahn and Meitner we thought that they were better prepared than us for this kind of chemical work and abandoned the study of the 'uranium puzzle' for concentrating our efforts on other themes...Hahn,

Meitner, and Strassmann went on for years following the same lines in the experimental work and in the interpretation of their results [56].

With the focus of Fermi's group shifted elsewhere, their primary rivals in the investigation of the transuranes was a group led by Irène Curie, who herself split the Nobel Prize for Chemistry with her husband in 1935.

Meitner and Hahn extended Fermi's results, demonstrating that two of the products of bombardment of uranium by neutrons were not consistent with elements from $Z = 80$ to $Z = 92$. But they and Curie's group soon found not just two transuranes, but a bewildering array of at least five different transuranic elements, each with multiple isotopes [57]. This was at the time unprecedented among the elements, although Meitner and Hahn soon found that bombarding thorium yielded a similarly puzzling array of products [11].

Meitner's role in the collaboration did not cease when she fled to Sweden. She and Hahn continued to exchange letters, often writing several in the span of a week. While some of those letters dealt with the aftermath of Meitner's flight, personal news, or administrative matters, they also continued to discuss the results of the ongoing experiments on the transuranes.

One product, initially discovered by Curie's group, was particularly mysterious. Curie's group didn't know what to make of it: it behaved chemically somewhat like lanthanum ($Z = 57$). Actinium ($Z = 89$) behaves chemically like lanthanum, and would thus be a plausible element with an atomic number close to uranium, but Curie's group could demonstrate it was not actinium. That left them with the hypothesis that it was yet another transurane isotope. Since the chemical properties of the transuranes were not yet known, they provided a convenient option for any new substance which could not be identified with the elements leading up to uranium. But the number of such identifications was climbing alarmingly, and their relationships to each other were not clear. Hahn and his co-workers, particularly Strassmann, began to investigate this new substance.

At first, they thought the substance might be an isotope of radium—before long, they revised that to *three* isotopes of radium [12]. Meitner and Hahn managed to meet and talk it over, for a day, in Copenhagen, a meeting that they both had to keep secret lest Hahn be accused of conspiring with a disloyal Jew [11].

The exchange of letters continued. Finally, on 19 December, Hahn wrote to Meitner of the most puzzling result of all: the 'radium' isotopes appeared, by every chemical test, to actually be barium ($Z = 56$). It appeared that the uranium nucleus had somehow split into pieces [12].

It was not that no-one had thought of the possibility before; Noddack, for instance had included it in her paper in 1934. But it was, from a physicist's point of view, thought to be physically impossible (see the Science Summary at the end of this chapter for details).

Meitner was, in the words of her physicist nephew Otto Frisch, 'brooding over' Hahn's letter when Frisch came to visit her for Christmas [57]. After having Christmas dinner together the first night, the two physicists went for a walk in the woods the next day[8] to talk through the implications. The details of their discussion are covered in the Science Summary at the end of this chapter, but by the time they came back to their hotel, they had concluded that the splitting of a nucleus of uranium into two large chunks, plus some loose neutrons, was not only allowed by physics but quantitatively consistent with known characteristics of the nucleus.

Frisch returned to Copenhagen, so that the two of them composed their paper on the new phenomenon (dubbed 'nuclear fission' by Frisch) over the telephone [54]. He also conducted an experiment to directly detect the fission fragments as they formed, consulting closely with Meitner on the experimental design. Frisch sent the two papers, one a short theoretical note with Meitner as lead author and Frisch listed second [58], the other a brief description of Frisch's experimental results [59], to *Nature* on 16 January, leading to publication of the theory paper on 11 February and the experimental one on 18 February.

This achievement was bittersweet for Meitner. She and Frisch were the first to understand how nuclear fission occurred, and that understanding immediately led to the realization that experiments since Fermi's work of 1934 had been inducing fission: fission had been 'discovered.' But it also meant that all of her work with Hahn in identifying 'transuranes' was following a blind alley; nearly all of their transuranes were actually fission fragments [28].

In addition, their results meant that Fermi had been given a Nobel Prize based in part on an incorrect conclusion! His lecture had been delivered on 12 December, just weeks before the work of Meitner and Frisch made much of it moot. In it, he credited Hahn and Meitner for identifying four transuranes:

> It is known that O Hahn and L Meitner have investigated very carefully and extensively the decay products of irradiated uranium, and were able to trace among them elements up to atomic number 96 [60].

Prior to publication of the remarks he had given, Fermi added a footnote correcting this:

> The discovery by Hahn and Strassmann of barium among the disintegration products of bombarded uranium, as a consequence of a process in which uranium splits into two approximately equal parts, makes it necessary to reexamine all the problems of the transuranic elements, as many of them might be found to be products of a splitting of uranium.

Scientists often make mistakes, and the best of them freely admit those mistakes and move on. But this was still a bit embarrassing. The Swedish Academy did not

---

[8] Sime, in her biography of Meitner, thinks it more likely the date was the 24th. McGrayne, for her part, places it on the 30th. Depending on the vagaries of Frisch's memory, it's conceivable the date was anywhere within that range. Frisch has told the story in multiple publications, but never provides the date, leaving us to judge from contextual clues.

like to award Nobel Prizes until they were confident they would stand up, and Fermi's was outdated almost as soon as it was awarded. Fermi's footnote was unfortunate as well, as it made it sound like Meitner was in on the mistake, but was not responsible for the correction, a misconception which Meitner was worried about even before her paper with Frisch was published [11].

On the other hand, Meitner and Frisch's accomplishment in explaining fission was remarkable, particularly since both were experimentalists, and their joint paper was pure theory. It is also notable that Meitner was a refugee who had only been in Sweden for a few months, and that she had just turned 60.

There was little confusion among the press reports that year, however. A *New York Times* article from January, for example, captures Meitner's role and its importance:

> [Fission] work was continued at the Kaiser Wilhelm Research Institute for Chemistry at Berlin-Dahlem, Germany, by Dr Lise Meitner and Professor Otto Hahn, who had been working together for many years. Dr Meitner was discharged last year for racial reasons and she went to Stockholm, Sweden... Professor Hahn and Dr Strassmann reported their startling observation [regarding the barium product] on Jan. 6 without offering any theory to explain the new phenomenon. Never before had it been observed, or even suspected, that an element so far removed on the periodic table (uranium occupies No. 92 on the Periodic Table of the Elements, while barium occupies No. 56) and so much lighter could be created from another element so much heavier...The exiled Dr Meitner, in Stockholm, was continuing this work in collaboration with Dr R Frisch...When the work of their German colleagues came to their attention they came to the conclusion that they were here dealing with a new atomic process. They were the first to realize what was happening was the actual splitting of the uranium atom... [61]

In short, the record is clear. The sequence of events involved in the discovery of fission can be summarized as:

1. The physicist Rutherford, a Nobel Laureate, demonstrates induced transmutation of elements in 1919.
2. The physicist Joliot and the chemist Curie demonstrate the transmutation of a stable element in to a radioactive isotope of a different element in 1934. (They are awarded the Nobel Prize in Chemistry for this achievement in 1935.)
3. The physicist Fermi induces the fission of uranium in 1934, but mistakenly believes the fission products are transuranic elements. (He is awarded the Nobel Prize in Physics for this achievement in 1938.)
4. The physicist Meitner asks the chemist Hahn to join her in an investigation of the products of the bombardment of uranium, a project they pursue, along with chemist Strassmann, for the next four years.
5. Meitner flees to Sweden, but continues to work informally with Hahn via letters.
6. Hahn and Strassmann discover one of the products of uranium bombardment is barium, suggesting the nucleus may have split. (Hahn is awarded the Nobel Prize in Chemistry for this achievement in 1944, deferred to 1945.)

7. Meitner and the physicist Frisch come up with a theoretical explanation of the fission of uranium.
8. Frisch, consulting closely with Meitner, confirms the fission of uranium.

3.7 The breakdown of science

Looking at the sequence of events enumerated above, something seems to be amiss. It would have been natural for Hahn and Strassmann to receive a Nobel Prize in Chemistry for discovering the uranium nucleus had split, and for a Nobel Prize in Physics to be awarded to Meitner and Frisch for explaining how that could take place. And yet only Hahn emerged a laureate. What went wrong?

While much has been written on the subject, the short answer is that the practice of science as it had been established in earlier decades was breaking down. Hahn, still working in Nazi Germany, could not, if he wished to remain employed, give too much credit to Meitner, and certainly he could not admit that they had continued to collaborate after she fled. Meitner, in turn, could not discuss their collaboration without risking harm to Hahn. Meanwhile, the Allies, realizing the potential to develop an atomic bomb, began a campaign of self-censorship that left further development of fission out of the open scientific literature and out of the public eye. Germans were forbidden to accept Nobel Prizes [62], and the Swedish Academy was losing touch with much of the international scientific community. The process of attributing credit for discoveries, which is always fraught, acquired additional political charge.

And the Nobel process itself was sputtering: no Nobel Prizes in any categories were awarded for the years 1940–42.

Between 1938 and 1945, Hahn was nominated for the chemistry prize eight times (three of them jointly with Meitner), while Meitner was nominated for physics six times in the same span (five of them jointly with Hahn). But Hahn netted *ten* nominations for physics during those years, even though he was not a physicist and his role in the discovery of fission involved little physics.

One of the nominations is particularly revealing. Manne Siegbahn, a member of the Nobel Committee for Physics and the head of the institute of experimental physics in which Meitner now worked, nominated Hahn alone for the physics prize in 1943.

It is difficult to see Siegbahn's nomination of Hahn for the physics prize as motivated by sound scientific judgment. Meitner worked in his Institute, and thus there could have been a degree of personal antipathy involved. It's possible he might have been worried that if she won the Nobel Prize and the money that went with it that she could assert her independence and become an alternative academic power center at his institute. Or he might simply have seen her as unproductive and unimpressive—what Meitner saw as a lack of support and resources from Siegbahn he may have seen as a lack of initiative and independence from her. Perhaps a more important factor was how closely involved Siegbahn was in the scientific politics of Sweden.

For example, the 1943 Nobel Prize in chemistry was awarded to George de Hevesy, a Jewish scientist fleeing Denmark for a position at the University of Stockholm, where Siegbahn was a faculty member. In 1943, Hahn was also nominated for chemistry by a member of the Nobel Committee for Chemistry. Siegbahn's nomination for Hahn in physics may have been meant to suggest an alternative method for awarding a prize to Hahn, thus clearing the stage for de Hevesy. Or it might have been meant to assert that nuclear fission belonged to physics, and not to chemistry. Either way, adding Meitner and Frisch in to the mix could have made it less likely that those inclined to vote for Hahn in chemistry could be convinced to vote for him in physics instead.

As it turned out, the Academy did not make any choice for the chemistry or physics prizes in 1943, deferring their decision until 1944 and then retroactively bestowing the prizes on de Hevesy for chemistry [63] and Stern for physics [64].

It is unlikely we will ever fully sort out the mixture of influences on the Nobel Prizes awarded during the last years of World War II. Many different factors have been suggested as being in play, and it is exceedingly difficult to objectively evaluate their relative importance. Regardless, the nomination of Hahn alone for a prize in physics suggests that more than scientific merit was at play.

The chemistry prize did not make its way to Hahn easily. In 1943, the Nobel Committee for Physics rejected Siegbahn's claim that nuclear fission 'lies on the boundaries between physics and chemistry' and passed Hahn's nomination on to the Committee for Chemistry. The Committee for Chemistry then recommended Hahn for the chemistry prize in 1944, only to be voted down by the chemistry section of the Academy and the Academy as a whole, which decided to defer the decision until 1945. In 1945, the roles reversed, with the Committee for Chemistry recommending to defer awarding a prize for another year and the Academy as a whole narrowly awarding that prize to Hahn [65].

This sequence of events seemed to establish that fission was not a matter for the Committee for Physics, shutting Meitner out of consideration for the prize.

Of course, that judgment could have changed over time, and Meitner's supporters continued to nominate her, providing her an impressive 21 nominations for the physics prize after 1945, including ones from such luminaries as Niels Bohr, Max Planck, and, in 1948, Otto Hahn.

Unfortunately for Meitner, Hahn's support for her role outside the public eye, in the form of the 1948 nomination, was undermined by his public statements and those of his supporters [35]. Hahn downplayed Meitner's role in the discovery, even going so far as to suggest that her influence had kept fission from being discovered prior to 1939! When others referred to her as his assistant, he did not correct them. When the Deutsches Museum obtained a lab bench and equipment meant to commemorate the discovery of fission, it was labeled as Hahn's (figure 3.3). Replicas of the display were created for other museums, and shown around the world. Eventually, Hahn made an audiotape to accompany the display, which included no mention of Meitner. The injustices in the display and tape were not fully corrected until 1990 [66].

The degree to which these slights were conscious is open to interpretation, although they can not have been completely innocent. By all accounts Hahn was not an introspective man, and when, after the war, he was lionized by his countrymen as an example of a great scientist who was opposed to the Nazis and yet remained with his country, he may have become convinced of his own legend [35]. Perhaps he even

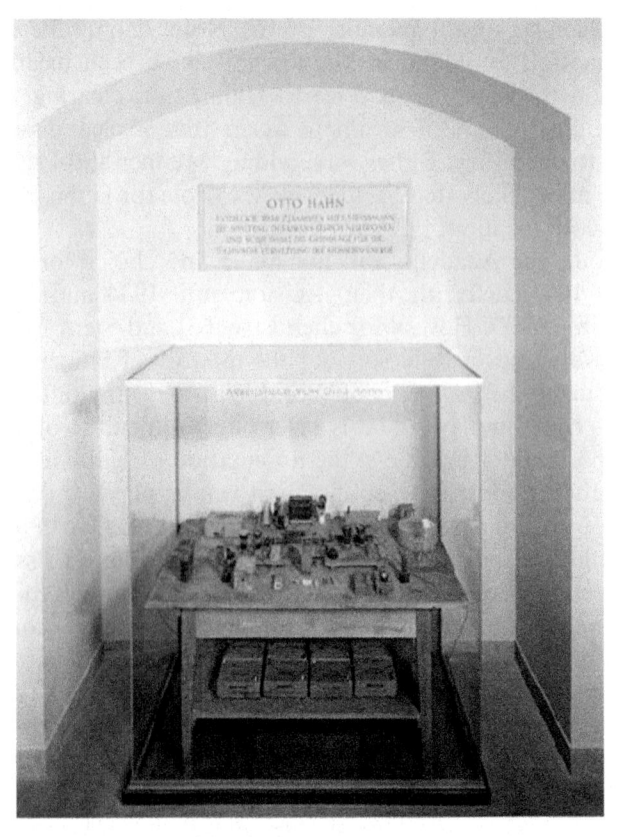

Figure 3.3. Nuclear fission display in the Deutsches Museum as it originally appeared, prior to the 1990 corrections. © Deutsches Museum, Munich, Archiv, BN30452.

suffered from a degree of impostor syndrome, and consciously or subconsciously feared that crediting Meitner forthrightly and publicly would reveal his own stature to be a fraud.

Whatever his reasons, Hahn's rhetoric muddied the waters. Meitner did develop a new research group in Sweden and continued to publish new scientific research into her 70s and articles on the history of science into her 80s. Over her career she received an array of honors including the Silver Leibniz Medal, the Ignaz Lieben Prize, the Ellen Richards Prize, the National Women's Press Club Woman of the Year, the Max Planck medal, the Otto Hahn Prize, the William Exner Medal, the Dorothea Schlözer Medal [55], and the Enrico Fermi Prize [12]. But she never received her Nobel.

3.8 Our Madame Curie

Einstein often referred to Meitner as 'our Marie Curie' [5]. While it is common for any successful female physicist to be compared to Curie, the parallels between Meitner and Curie, from the trivial to the profound, go further. They shared a birthday (7 November); they both had a delayed entrance to university, beginning at the age of 23; they both were experimental nuclear physicists; they both served as x-ray technicians during World War I, they both left their native country to do their best work; they both headed prominent scientific laboratories.

And while Marie Curie has two Nobel Prizes and Meitner none, there is one more similarity between them. For nuclear chemists and physicists, there is an honor even more rare than the Nobel Prize: having an element named after them [67]. In 1948, the transuranic element with atomic number 96 was named after Marie and Pierre Curie, becoming curium. Even heavier elements were named after Einstein, Fermi, Mendeleev, Nobel, and Lawrence. For elements 104, 105, and 106, however, disputes arose between American and Soviet teams as to who had discovered those elements first, and thus who had the right to suggest names for them. The Americans proposed hahnium for element 105, after Otto Hahn, but the Soviets preferred nielsbohrium. Meanwhile, a German group discovered elements 107, 108, and 109. The International Union of Pure and Applied Chemistry (IUPAC) offered a variety of compromises, including some which used some of the names suggested by the Americans and the Soviets for elements 104–106 for the elements discovered by the German group. Thus, at various times, hahnium was also suggested for elements 108 and 109. In the end, hahnium did not become the name of any element. The name the German group proposed for element 109, however, did end up becoming official in 1997.

Element 109, a transurane, is now known as meitnerium.

3.9 Science summary: nuclear fission

To understand why nuclear fission was first thought to be impossible, we first need to understand the mechanism of alpha decay, in which an unstable nucleus spontaneously emits an alpha particle. For specificity, let's consider the alpha decay of ^{238}U, the most abundant isotope of uranium.

Alpha particles consist of two protons and two neutrons, and thus have a charge of +2 fundamental units. Removing this unit from ^{238}U would reduce the atomic number (number of protons) by two and the mass number (number of protons + neutrons) by four, resulting in a *daughter nucleus* of ^{234}Th. As soon as the positively charged alpha particle is emitted, it is repelled forcefully by the positively charged daughter nucleus. According to Coulomb's Law, the amount of energy of two nearby charged objects is proportional to the product of the charges and inversely proportional to the distance between them.

Because nuclei with a positive charge are a feature of our universe, it must be the case that when the protons and neutrons are brought very close together that another force comes into play, attractive rather than repulsive and much stronger than the ordinary Coulomb force between charged particles. Physicists, employing a particularly prosaic turn of phrase, call this strong force that acts only at very short distances the *strong force*.

Imagine the reverse process of alpha decay, in which an alpha particle is shot at a nucleus of ^{234}Th. As the alpha particle approaches the nucleus, the energy of the interaction of the combination would become greater and greater. In practice, this energy would come at the expense of the kinetic energy (energy of motion) of the

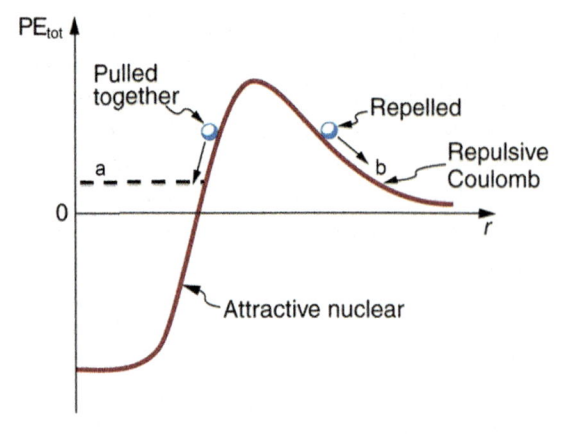

Figure 3.4. Schematic of interaction energy as a function of the distance r between an alpha particle and a nucleus. (Modified version of figure from OpenStax College, Physics. OpenStax CNX. Download for free at http://cnx.org/contents/031da8d3-b525-429c-80cf-6c8ed997733a@9.77.).

incoming alpha particle. According to classical (19th century) physics, if the particle were not energetic enough to reach the distances where the strong force takes over, then the alpha particle would first slow to a stop and then convert the energy of action back in to kinetic energy, shooting off in a different direction. If, on the other hand, the alpha particle had enough energy to penetrate the nucleus, it would also have enough energy to escape it. Capture could occur only if some of the energy were dissipated, perhaps first by spreading it among the other protons and neutrons in the nucleus, and then by releasing it in the form of a gamma ray photon. A graph of the energy of interaction versus the distance of the alpha particle from the nucleus would look schematically like figure 3.4.

Now let's consider the process of alpha decay in the forward direction, where we start with a nucleus of ^{238}U and end with a nucleus of ^{234}Th and an ejected alpha particle.

Initially, the alpha particle would be within the nucleus, and thus in the left section of figure 3.4. Because ^{238}U is unstable relative to alpha decay, the energy of interaction would be greater than if the particles were widely separated. This energy is shown by the dotted line marked a in the figure. While the nucleus is unstable, it does not disintegrate immediately, because there is not enough energy to get the alpha particle over the peak in figure 3.4, known as the *Coulomb barrier*. Classically, the only way the alpha particle could get out is if energy were added to the system, perhaps from an incoming gamma ray or through collision with another particle.

Quantum mechanics, however, allows for another possibility. Positions in quantum mechanics are always a little uncertain. It's possible for the alpha particle to, in essence, 'forget' which side of the Coulomb barrier it was on; this is called *quantum tunneling*. If it were to find itself at the position marked b on the graph, it would be rapidly repelled and ejected from the nucleus. This is the mechanism of radioactive decay.

An alternative, but equally valid, way of thinking about this process is to say that, under quantum mechanics, the energy of a system is not well defined over short time periods. For a short span of time, for example, the system might have enough energy to get over the Coulomb barrier. To understand this way of thinking about it, consider the analogy of someone who likes to pay his credit cards off in full at the end of each month. One day, he discovers he didn't bring enough cash with him to buy a train ticket home from work, although he has money at home. So he uses his credit card to buy a train ticket home. The next day, he takes the money he has at home and sends it to the credit card company, paying off his bill. If he wasn't able to borrow money using their credit card, he would have been trapped at work until he got paid; similarly, under classical physics the alpha particle would be trapped in the nucleus unless it got energy from outside. But with the credit card, he could penetrate through the 'barrier' of his commute, at which point he could pay back the debt.

Prior to the work of Meitner and Frisch, there were only two mechanisms that had been considered which would break a nucleus apart: providing enough energy to get over the Coulomb barrier, or using quantum tunneling to get through it.

When ^{238}U ($Z = 92$) decays into ^{234}Th ($Z = 90$) and an alpha particle ($Z = 2$), the product of the charges is $(+2)(+90)$, or $+180$ in atomic units. But if ^{238}U were to 'decay' into barium ($Z = 56$) and krypton ($Z = 36$), the product of the charges would be $(+56)(+36) = +2016$, more than ten times higher. Since the Coulomb barrier is proportional to the product of the two charges, it would also be on the order of ten times higher. In addition, according to basic quantum theory, the probability of tunneling is lower for higher mass particles. The krypton nucleus that would be 'ejected' in the second 'decay' is roughly 20 times heavier than an alpha particle. On the other hand, the energy released in fission would also be much higher; i.e. the nucleus is more unstable relative to ejection of a krypton nucleus than an alpha particle. In summary, compared to ejecting an alpha particle, the ejection of a krypton nucleus requires tunneling through a higher Coulomb barrier with a heavier particle that is thus less prone to tunneling, but the final result of the process would be more energetically favored. Which effect wins out requires some analysis [68], but the net result is that the spontaneous fission of uranium via tunneling would be much less probable than the ejection of an alpha particle. To return to the analogy of the worker without train fare, it is as if, in addition to the small amount of cash he has at home, he also has a large amount of cash at his parents' house on the other side of the country. But, while using his credit card to fly there would indeed result in him having more spending money than just using his credit card to get home, it involves much more trouble than his short commute, so he chooses the commute.

While the neutron bombardment experiments which led to induced fission did involve some addition of energy from the incoming neutron, calculations showed it wasn't nearly enough to favor fission over other modes such as alpha emission. To make matters worse, some products were best formed with bombardment by *slow* neutrons, for which the added energy was negligible.

And yet, according to the experiments of Hahn and Strassmann, fission had occurred, with one of the products being barium. If a uranium nucleus had split in two and one daughter was barium, then conservation of charge required the other to be krypton. If there was not enough energy in the bombarding neutron to surmount

Figure 3.5. A liquid drop distorted into two lobes.

the Coulomb barrier, and if the krypton nucleus was too heavy and too highly charged to tunnel through it, then what else could have been happening?

Up until that time, physicists who had briefly considered and then rejected the possibility of fission had thought of atomic nuclei either as solid objects, like little chunks of crystal [28], or as miniature solar systems, with nuclear units orbiting around each other. But those weren't the only models in use for the nucleus.

George Gamow had developed a model of the nucleus in which it behaved much like a liquid drop. This model was later extended by others, including Heisenberg and Bohr [69].

If a nucleus was like a liquid drop, then it could behave collectively, changing shape. During their walk in the woods, Frisch computed that the 92 units of positive charge in the uranium nucleus were enough to almost completely cancel the surface tension of the nuclear drop, allowing it to deform and take on non-spherical shapes with ease [57]. The addition of just one captured neutron, therefore, might be enough to distort it into a shape such as is shown in figure 3.5, with two distinct portions of the nucleus separated by a narrow waist [69].

A gradual rearrangement of the nucleus into this shape, with charges flowing as necessary to facilitate the distortion, would reduce the Coulomb barrier to the point where it could split classically, without the need for any tunneling at all. Physicists had been thinking of the Coulomb barrier to nuclear disintegration as fixed, a mountain which an ejected particle needed to go over or through. But Meitner and Frisch realized it was more like an enormous wave on the open ocean—an enterprising surfer could go over it, a brave swimmer could go through it, but another possibility was for the wave itself to dissipate, allowing easy passage.

On that same walk, Meitner then worked out the repulsive energy of the two drops—the nuclear fragments—at the moment of separation, and compared to the known difference in mass between the daughters (krypton and barium) and the parent (uranium). Using Einstein's formula $E = mc^2$, she found that the energy of repulsion agreed with the difference in mass [69]. This allowed her to predict the kinetic energy of the ejected nuclei, which was then confirmed by Frisch's measurements when he returned to Copenhagen.

After years of confusion, Meitner and Frisch used a mixture of classical physics, quantum mechanics, and relativity to make everything clear—physicists around the world understood immediately, and rushed to investigate and extend this new understanding. And the world was changed forever.

References

[1] Wright H 1949 *Sweeper in the Sky: The Life of Maria Mitchell, First Woman Astronomer in America* (New York: Macmillan)
[2] Kelly S E and Rosner S A 2012 Winifred Edgerton Merrill: 'She Opened the Door' *Not. AMS* **59** 504–12
[3] Creese M R S 2004 *Ladies in the Laboratory II: West European Women in Science, 1800–1900: a Survey of Their Contributions to Research* (Lanhan, MD: Scarecrow)
[4] Women's Higher Education Institution (Bestuzhev Courses) Opened in St Petersburg *Yeltsin Presidential Library* accessed 8 April, 2017. http://www.prlib.ru/en-us/History/Pages/Item.aspx?itemid=679
[5] Lewin Sime R 2005 From exceptional prominence to prominent exception: Lise Meitner at the Kaiser Wilhelm Institute for Chemistry *Vorabdrucke aus dem Forschungsprogramm 'Geschichte der Kaiser-Wilhelm-Gesellschaft im Nationalsozialismus' Ergebnisse* **24**
[6] Curie E 1937 *Madame Curie* (Garden City, NY: Doubleday, Doran & Company)
[7] Clark L L 2008 *Women and Achievment in Nineteenth-Century Europe* (Cambridge: Cambridge University Press)
[8] Ehrenhaft-Steindler O *LISE: Naturwissenschaften Unterricht Mädchen* accessed 8 April, 2017 http://lise.univie.ac.at/physikerinnen/historisch/olga_ehrenhaft-steindler.htm
[9] Rentetzi M 2009 *Trafficking Materials and Gendered Experimental Practices: Radium Research in Early 20th Century Vienna* (New York: Columbia University Press)
[10] Meitner L 1964 Looking back *bull. Atom. Sci.* **20** 2–7
[11] Lewin Sime R 1996 *Lise Meitner: A Life in Physics* (Berkeley, CA: University of California Press)
[12] Frisch O R 1970 Lise Meitner: 1878–1968 *Biographical Memoirs of Fellows of the Royal Society* **16** 405–20
[13] Krafft F 1983 Internal and external conditions for the discovery of nuclear fission by the Berlin team *Otto Hahn and the Rise of Nuclear Physics* ed W R Shea (Dordrecht: Reidel)
[14] Hahn O 1966 *Otto Hahn: A Scientific Autobiography* (trans) ed Willy Ley (New York: Scribner)
[15] Bertsch McGrayne S 1998 *Nobel Prize Women in Science: Their Lives, Struggles, and Momentous Discoveries* 2nd edn (Washington: Joseph Henry)
[16] Rasetti F 2003 interview by Judith R Goodstein *Oral History Project of the California Institute of Technology Archives* http://oralhistories.library.caltech.edu/70/1/OH_Rasetti.pdf
[17] Kuhn T S 1970 *The Structure of Scientific Revolutions* 2nd edn (Chicago, IL: University of Chicago Press)
[18] Frisch O 1978 Lise Meitner, nuclear pioneer *New Scientist* November 9, 1978
[19] Hahn O and Meitner L 1909 Eine neue Methode zur Herstellung radioaktiver Zerfallsprodukte; Th D, ein kurzlebiges Produkt des Thoriums *Verh. D. Phys. Ges.* **11** 55–61
[20] Fry C and Thoennessen M 2013 Discovery of the thallium, lead, bismuth, and polonium isotopes *Atom. Data Nucl. Data* **99** 365–89

[21] Fry C and Thoennessen M 2013 Discovery of the actinium, thorium, protactinium, and uranium isotopes *Atom. Data Nucl. Data* **99** 345–64
[22] Hahn O and Meitner L 1909 Nachweis der komplexen Natur von Radium C *Phys. Z* **10** 697–703
[23] Duparc O H 2009 Pierre Auger—Lise Meitner: comparative contributions to the Auger effect *Inter. J. Mater. Res.* **100** 1162–6
[24] Andrade E N da C 1926 *The Structure of the Atom* 3rd edn (New York: Harcourt, Brace, & Co)
[25] Leone M and Robotti N 2010 Fréderic Joliot, Irène Curie and the early history of the positron (1932–33) *Eur. J. Phys.* **31** 975–87
[26] Hentschel K (ed) 1996 Letter from Lise Meitner to Otto Hahn, March 21, 1933 *Physics and National Socialism: An Anthology of Primary Sources* (trans) Ann M Hentschel (Basel: Birkhäuser Verlag)
[27] Frisch O R 1979 *What Little I Remember* (Cambridge: Cambridge University Press)
[28] Interview of O Frisch by Charles Weiner on May 3, 1967, Niels Bohr Library & Archives, American Institute of Physics, College Park, MD USA, www.aip.org/history-programs/niels-bohr-library/oral-histories/4616
[29] Hentschel K 1996 introduction to *Physics and National Socialism: An Anthology of Primary Sources* ed Klaus Hentschel (Basel: Birkhäuser)
[30] Hentschel K (ed) 1996 Law for the restoration of the professional civil service *Physics and National Socialism: An Anthology of Primary Sources* (trans) Ann M Hentschel (Basel: Birkhäuser)
[31] Hentschel K (ed) 1996 Third ordinance on the implementation of the law for the restoration of the professional civil service *Physics and National Socialism: An Anthology of Primary Sources* (trans) Ann M Hentschel (Basel: Birkhäuser)
[32] Ball P 2014 *Serving the Reich: the Struggle for the Soul of Physics under Hitler* (Chicago, IL: Chicago University Press)
[33] Lenard P and Stark J 1996 Hitlergeist und Wissenschaft *Großdeutsche Zeitung* May 8, 1924 1-2 (translated as *The Hitler Spirit and Science* in *Physics and National Socialism: An Anthology of Primary Sources* ed Klaus Hentschel (trans) Ann M Hentschel (Basel: Birkhäuser)
[34] Kentschel K (ed) 1996 'Weiße Juden' in der Wissenschaft *Das Schwarze Korps* July 15, 1937, 6 (translated as *'White Jews' in Science* in *Physics and National Socialism: An Anthology of Primary Sources* (trans) Ann M Hentschel (Basel: Birkhäuser)
[35] Lewin Sime R 2006 The politics of memory: Otto Hahn and the Third Reich *Phys. Perspect.* **8** 3–51
[36] LeBor A and Boyes R 2001 *Seduced by Hitler: The Choices of a Nation and the Ethics of Survival* (Naperville, Illinois: Sourcebooks)
[37] Majer D 2003 *'Non-Germans' Under the Third Reich: The Nazi Judicial and Administrative System in Germany and Occupied Eastern Europe, with Special Regard to Occupied Poland, 1939-1945* (trans) Peter Thomas Hill, Edward Vance Humphrey, and Brian Levin (Baltimore: Johns Hopkins University Press)
[38] Fermi Weiner N 2009 'Biography of Laura Fermi' *The Fermi Effect: A Living Legacy Project*, accessed April 8, 2017. http://fermieffect.com/biography-of-laura-fermi
[39] Pearl Buck Wins Nobel Literature Prize; Third American to Get the Swedish Award *New York Times (1923–Current File)* November 11, 1938 1

[40] History *Kungl. Vetenskaps Akademien*, accessed April 8, 2017. http://www.kva.se/en/About-the-academy/History
[41] Royal Swedish Academy of Sciences *IAP: The Global Network of Science Academies*, accessed April 8, 2017. http://www.interacademies.net/Academies/ByRegion/EuropeCentralAsia/Sweden/12506.aspx
[42] Members: Physics *Kungl. Vetenskaps Akademien*, accessed February 19, 2017. http://www.kva.se/en/contact/Members/Physics/
[43] MacLachlan J 1991 Defining physics: The Nobel Prize selection process, 1901–1937 *Am. J. Phys.* **59** 166–74
[44] Nobel A B 2014 Will, translated as *Full Text of Alfred Nobel's Will* Nobel Media AB 2014 (web) 9 April 2017. http://www.nobelprize.org/alfred_nobel/will/will-full.html
[45] Crawford E and Lewin Sime R 1997 A Nobel tale of postwar injustice *Phys. Today* **50** 26–32
[46] Nomination and Selection of Physics Laureates *Nomination* accessed April 8, 2017 https://www.nobelprize.org/nomination/physics/index.html
[47] Lise Meitner *Nomination Database* accessed February 19, 2017 https://www.nobelprize.org/nomination/archive/show_people.php?id=6097
[48] Joliot F and Curie I 1934 Artificial production of a new kind of radio-element *Nature* **133** 201–2
[49] Rutherford E 1919 Collisions of alpha particles with light atoms. IV. An anomalous effect in nitrogen *London, Edinburgh and Dublin Phil. Mag. and J. Sci.* **37** 581–7
[50] Element 93 *New York Times (1923–Current File)* June 6, 1934, 20
[51] Ida Noddack *Nomination Database* accessed February 19, 2017 https://www.nobelprize.org/nomination/archive/show_people.php?id=6731
[52] Noddack I 1934 über das Element 93 *Angewandte Chemie* **47** 653–5 (trans); Graetzer H G *On Element 93* accessed April 8, 2017 http://bourabai.kz/tyapkin/element93.htm
[53] The Nobel Prize in Physics 1938 *Nobel Prizes and Laureates* accessed April 8, 2017 https://www.nobelprize.org/nobel_prizes/physics/laureates/1938
[54] Meitner L 1962 Right and wrong roads to the discovery of nuclear energy *IAEA Bull.* December 2 6–8
[55] Ogilviy M and Harvey J 2000 *The Biographical Dictionary of Women in Science: Pioneering Lives from Ancient Times to the Mid-20th Century* (New York: Routledge)
[56] Amaldi E 1984 Neutron work in Rome in 1934-36 and the discovery of uranium fission *Riv. Stor. Sci.* **1** 1-24 (reprinted 1998 *20th Century Physics: Essays and Recollections: A Selection of Historical Writings by Edoardo Amaldi* ed Giovanni Battimelli and Giovanni Paoloni (Singapore: World Scientific))
[57] Frisch O R and Wheeler J A 1967 The discovery of fission *Phys. Today* **20** 43–52
[58] Meitner L and Frisch O R 1939 Disintegration of uranium by neutrons: a new type of nuclear reaction *Nature* **143** 239–40
[59] Frisch O R 1939 Physical evidence for the division of heavy nuclei under neutron bombardment *Nature* **143** 276
[60] Fermi E 1965 Artificial radioactivity produced by neutron bombardment *Nobel Lectures Physics 1922-1941* (Amsterdam: Elsevier)
[61] Vast energy freed by uranium atom *New York Times (1923–Current File)* January 31, 1939, 18
[62] Germany and the Nobel Prizes 1937 *Science* **85** 171

[63] The Nobel Prize in Chemistry 1943 *Nobel Prizes and Laureates*, accessed April 8, 2017 http://www.nobelprize.org/nobel_prizes/chemistry/laureates/1943
[64] The Nobel Prize in Physics 1943 *Nobel Prizes and Laureates*, accessed April 8, 2017 http://www.nobelprize.org/nobel_prizes/physics/laureates/1943
[65] Crawford E, Lewin Sime R and Walker M 1996 A Nobel tale of wartime injustice *Nature* **382** 393–5
[66] Lewin Sime R 2010 An inconvenient history: the nuclear-fission display in the Deutsches Museum *Phys. Perspec.* **12** 190–218
[67] Stuart Fox 2009 What's it like to name an element on the periodic table? *Pop. Sci.* June 26 http://www.popsci.com/scitech/article/2009-06/whats-it-name-element-periodic-table
[68] Hübsch T 2012 The theory of alpha decay, accessed June 17, 2017, http://physics1.howard.edu/~thubsch/QM2/AlphaDecay.pdf
[69] Stuewer R H 1997 Gamow, alpha decay, and the liquid-drop model of the nucleus *George Gamow Symposium, ASP Conference Series* **129** 29–43

IOP Concise Physics

Beyond Curie
Four women in physics and their remarkable discoveries, 1903 to 1963
Scott Calvin

Chapter 4

Chien-Shiung Wu

4.1 Mighty hero

In physics in 1940, no topic was hotter than nuclear fission, and the University of California, Berkeley was a rising powerhouse in the field. Faculty included Ernest Lawrence, winner of the Nobel Prize in physics in 1939 for the invention of the cyclotron, who would eventually be honored by having his name bestowed on element 103; J Robert Oppenheimer, who would take a leave a few years later to direct the American atomic bomb effort at Los Alamos; Luis Alvarez and Emilio Segrè, future Nobel laureates in physics; and the chemist Glenn Seaborg, who would earn both a Nobel in Chemistry and a place on the periodic table. Postdocs included Julian Schwinger, another future Nobel laureate in physics, and Chien-Shiung Wu (吳健雄), who had just received her doctorate under the supervision of Lawrence, Oppenheimer, and Segrè [1]. Wu, following on the excitement generated by Meitner and Frisch's interpretation of Hahn and Strassmann's results, was studying the products of uranium fission, and identified among them two radioactive isotopes of xenon, including the previously unknown isotope xenon-135 [2, 3]. This was just the beginning of her remarkable career.

Wu was born in Shanghai in the Spring of 1912 [4]. Chinese given names are not as gendered as western ones. Many names are gender-neutral, and even those that are associated with a particular gender are still sometimes given to babies of the other gender [5]. Chien-Shiung means 'mighty hero,' both halves of which (each one character in Chinese script) are traditionally considered masculine names in Chinese culture[1]. Chien was a generational name in Wu's family, so that her siblings also had it as a character in their name. In traditional Chinese culture, these generational names usually follow an algorithmic pattern, such as being sequential words in a poem [6]. In Wu's family, it appears the remaining character of their names was also

---

[1] A simple test: www.chinese-tools.com/tools/gender-guesser.html identifies both names as likely male.

assigned algorithmically. Thus, the second child in Wu's family would have been named Chien-Shiung regardless of gender [4].

But while algorithmic naming was a traditional practice, applying the generated names without regard to gender was likely a deliberate break with tradition on the part of Wu's parents [6]. Wu's father was literally a revolutionary, participating actively in the revolutions of 1911 and 1913. His wife changed her name to take one with a revolutionary meaning (Fu Hua, 'Rebuild China') [4]; in this context it is not surprising that they gave their children, including Chien-Shiung, martial names as well.

Wu's father, in addition to fighting for revolution, was well read, cosmopolitan, and a supporter of women's rights [7]. Through most of the 19th century, it was very unusual for Chinese girls to receive any education outside of the home; a common saying was that 'a woman's virtue lies in her ignorance' [8]. In the early 20th century that was beginning to change, and Wu's father founded a small girls' school using the name Ming De, or 'understanding morality', meant to contrast with the old idea. Still, the innovation initially lay more in the school being physically outside of the home than in the choice of subject matter, which was dominated by domestic skills such as sewing and gardening [4].

By this time, however, change was coming rapidly to China. The Qing Dynasty fell the year Wu was born. While the ensuing period is known as the Republic of China, it was dominated by repeated rebellions and struggles between warlords. Still, social change moved forward as part of the New Culture Movement, which included a significant component of feminism.

By the time Wu was nine, therefore, she was able to travel to attend the Soochow Girls School, a highly selective boarding school. This school consisted of two sections, one a regular high school, and the other a teacher-training program which provided free room and board and a guaranteed job after graduation, and required graduates to teach for at least one year immediately after graduation. Wu decided to apply for the teacher-training program, in essence because it was harder to get into and thus more prestigious. While she was admitted into the program, her choice proved to be a mistake, as she was primarily interested in the science and language courses being offered in the regular portion of the school. She ended up borrowing the books for those classes from other students, and using those to teach herself at night. Apparently a bit of a Francophile, she admired both Marie Curie and Napoleon Bonaparte, whom she said showed 'genuine feeling and concern toward his army.' She never did teach the required year, instead fulfilling her responsibility by spending a year at the National China College in Shanghai, the first private university in China.

Wu also had the option of going to a public university, as she was also accepted into the National Central University in Nanjing, which she transferred to a year later. Public universities in China had been co-ed since 1920, when Wu was only eight years old. I have said earlier that Meitner might have had an easier time if she had been born just a few years later, but Wu was born at just the right time. A dozen years earlier, and she would have faced the same kinds of delays that Meitner encountered. A few years later, and war would have intervened.

While Wu was already drawn to physics, she worried that her stolen moments with science books in high school were not enough of a preparation for majoring in physics at a top university, and considered continuing with an education degree. But her father saw where her true interest lay, and brought home three books: one on algebra, one on chemistry, and one on physics, and she spent a summer in more self-study [7].

Wu was still not confident she was ready for physics, however, and spent her first year at Nanjing as a mathematics major. A year later, she finally switched to physics, and became an all-around excellent student, winning praise from faculty and students alike [4].

Remembering this fifty years later, Wu marveled at how the entire trajectory of her life depended on those three books from her father:

> Imagine what a near miss it had been. If it hadn't been for my father's encouragement, I wouldn't have had the courage to select physics as a major field and I would be teaching grade school somewhere in China now [7].

Even that scenario was optimistic. If 'Mighty Hero' Wu had gained an education degree and taken up teaching in China, she would have had to navigate a brutal invasion; wars world, civil, and cold; and revolutions communist and cultural. With her revolutionary roots, intellectual background, and outspoken nature, it would have been a minor miracle for her to emerge unscathed.

Indeed, by the early 1930s, student protests had been the norm for a generation in China, with the latest wave protesting a succession of embarrassments at the hands of the Japanese. Wu was recruited to help lead the student movement at Nanjing, and did so in a way that satisfied the students and yet avoided confrontations that could have spiralled out of control. One sit-in she helped lead took place at the Presidential Palace, resulting in a brief colloquy with President Chiang. Even had she chosen the profession of school teacher, it is difficult to picture her having laid low through the turmoil that was to come!

Wu graduated *summa cum laude*, and after a year serving as a teaching assistant at Zhejiang University and a research assistant at Academia Sinica (the national scientific academy of China) [9], she decided to travel to the United States to pursue a doctoral education at the University of Michigan. Her uncle covered the cost of the travel [4].

Wu chose the University of Michigan in part because her advisor at Academia Sinica, a physicist named Jing-Wei Gu, had herself got her doctorate from Michigan. In fact, Michigan had gained a reputation as being welcoming for Chinese students, with more than 600 attending at the time [10]. On the way, Wu planned to stop for a week or so in Berkeley, visiting a friend. From there, she would travel on to Michigan, secure her doctorate, and return to her family in China.

4.2 Exile

This plan did not survive the week.

First, she heard that Michigan had segregated student unions, the Michigan League for women [11] and the Michigan Union, which had just been expanded, for men [12]. While both were housed in elegant buildings, the men's space was much larger, and included a bowling alley, a swimming pool, and a billiards room, features not found in the women's club. Women were allowed to enter the men's building, but only if they were escorted by a man, and only through the side entrance. Wu, daughter of a revolutionary and herself a student activist, was disappointed to learn that the United States could itself be so socially backward [4]. Even so, she did not come to the United States in order to be surrounded by people from China, so that the large concentration of Chinese students at Michigan began to seem like a drawback to her.

But neither of those reasons would have been sufficient to keep her from Michigan if she did not have a more appealing choice in front of her. In her week at Berkeley, she saw well-outfitted labs stocked with advanced equipment, including Lawrence's 37-inch cyclotron. In Wu's words:

Berkeley was at the top of the world…Physicists all over the world came to visit the famous cyclotron laboratory. It was the mecca, the holy land for atom smashing [7].

In that first week, Wu met several interesting and friendly graduate students, among them Luke Yuan, her future husband. Their romance did not develop until years later [4], however, and there is no indication that 'love at first sight' played a role in Wu's attraction to Berkeley.

The faculty, too, were welcoming, although Segrè, in his autobiography, makes clear that the internal politics and diverse personalities of Berkeley physics could be tricky. To start with, there was the distinction between the physics department of the University, with Raymond Birge as its chairman, and the Rad Lab directed by Lawrence. Birge, according to Segrè, was 'in some respects…a narrow-minded man, with prejudices against foreigners, especially Chinese, women, and anyone who spoke with an accent', all of which applied to Wu. Oppenheimer was 'considered a demigod by himself and others at Berkeley, and as such he spake in learned and obscure fashion'; his group's 'physics was valid, but often they attacked problems prematurely, or problems beyond their capabilities, resulting in indifferent success'. Lawrence was focused on his cyclotrons to the exclusion of other science, at times was lacking in tact, and had 'intentions and capabilities [that] could change at any time'. Fermi, who occasionally visited Berkeley, 'doubted Lawrence knew or understood much physics, and thought he was rather full of himself'. Of course, these are Segrè's characterizations; he admits he himself could be a bit of a 'curmudgeon' [13].

Of Wu, however, Segrè had nothing but praise:

She was a fiend for work, almost obsessed by physics, highly talented, and very shrewd, as well as witty…I admired her and liked her, and we remained friends for life [13].

Birge, whatever his general attitude toward foreign Chinese women who spoke with accents, knew talent when he saw it, and did not hesitate to offer Wu immediate admission to the graduate program, even though the semester at Berkeley had already started.

Most young scientists in Wu's place would have navigated the powerful personalities of the senior scientists by aligning themselves with one or the other of the luminaries, trusting their mentors to shield them from the vicissitudes of academic politics. Wu took a different approach (figure 4.1). Lawrence became her formal thesis advisor, and did not regret it, later writing in an admiring but sexist letter of recommendation that she was 'the ablest woman physicist that I have ever known… and altogether a decorative addition to any laboratory' [14]. Segrè, himself only a visiting scientist (i.e. a Jewish refugee from fascist Italy without a permanent position) at the time, was her informal advisor; the two frequently collaborated on experiments. She took classes from Oppenheimer, and the two formed a close relationship. Years later, she would still refer to him by the familiar 'Oppie'; she thought of him as a 'brilliant' theorist, while he admired the quality of her experimental work [4]. And Birge, while not working directly with Wu on scientific projects, made sure she felt welcome and safe in her new home.

Wu would need all the friends and supporters she could get. Less than a year after her arrival in Berkeley, in 1937, increasingly extreme Japanese provocations finally became too much for President Chiang, triggering all out war between China and Japan. By the end of the year, Japan had captured Nanjing, and in the ensuing weeks murdered and raped tens of thousands of Chinese civilians.

Figure 4.1. Left to right: Segrè, Oppenheimer, and Wu. © 2010 The Regents of the University of California, through the Lawrence Berkeley National Laboratory.

Eventually, the war between China and Japan would become just one of the many theaters of operation in World War II. After Japan was defeated, the Chinese Civil War between President Chiang's nationalists and communist forces consumed the next four years. Wu's passport from the Republic of China was no longer valid in many countries, and so she changed her plans again, and became a US citizen in 1954. Cold War restrictions then prohibited her from travelling to communist China. She did not make it back to China until 1962, and then only to Chiang's island stronghold of Taiwan. By that time her father, mother, and elder brother had passed away, but she did have a bittersweet reunion with her uncle and younger brother. Finally in, 1973, now past the age of 60, she returned to mainland China and the town of her birth. By that time her uncle and remaining brother were also gone, tortured to death during the Cultural Revolution. Even the tombs of her parents had been destroyed [4].

4.3 Pushing back

As a physicist, Wu had a successful career. Wu understood the challenges sexism and racism presented for her, but she suffered less of the kind of lack of advancement that plagued Payne and Mayer in the middle of their careers, or the catastrophic ethnic prejudice that interrupted Meitner in hers. Rather than take this for granted, she spoke out. In 1964, for example, she participated in a conference on American Women in Science and Engineering, held at MIT. The keynote address was given by Professor Bruno Bettelheim, a pompous but prominent professor of psychology from the University of Chicago [15][2].

On the topic of women in the professions, Bettelheim had much to say [16], although he freely admitted that he knew 'little about the specifics of the sciences (other than the social sciences) or about requirements for employment in such fields as physics, mathematics, and engineering'. Some of what he claimed was false, some simply insensitive. A few excerpts from the statements he made during his address:

> Science and engineering have been the endeavors of men for so long that it is understandable that the first women who entered these fields were often women who in many ways felt more like men than the rest of their sex…
>
> In our work with severely disturbed children, we do not find it too difficult to find women committed to working for and with these psychotic young-sters…our problem lies with recruiting male workers…the trouble with most of them is that they are committed to do their work in too feminine a way. This just will not do…

---

[2] Since his death, Bettelheim has become a very controversial figure. He is in some senses a charlatan, having parlayed claims of degrees and training he did not receive in to his position at Chicago. His belief that autism was caused by a lack of affection by the child's mother lacked evidence and has understandably angered many. He has also been accused of plagiarism, child abuse, and sexual harassment. Nevertheless, there are some who found his therapeutic methods useful, and still admire him as a great man. A good starting point to learn more about Bettelheim is the article in the New York Review of Books by his editor, which discusses five books with a variety of perspectives on Bettelheim.

While man's principal means of self-realization was through work, he could not achieve it without his role as husband and father becoming central to the meaning of his life. Woman's parallel path to autonomy lay in being wife and mother. But unless she, too, had a meaningful share in the work of preserving the present generation and of extending the horizons of future generations, procreation alone was not enough to fill out the meaning of her life. Before birth control the more numerous pregnancies certainly made her fully aware of her procreative role. That she was also worn out by these pregnancies as much as by physical or economic hardships is here beside the point...

To remedy this we must start with the realization that, as much as women want to be good scientists or engineers, they want first and foremost to be womanly companions of men and to be mothers...

Women feel more comfortable in dealing with problems of life that unfold within well-defined space; they are masters in exploring and fulfilling the requirements of living within an 'inner space' of experience. Men are equally drawn to mastery of external space (which only incidentally includes stellar space but pertains chiefly to mastery over nature at large).

Wu could not have been happy hearing these sentiments. For her part, she and Luke had one child, Vincent. Their care for Vincent included a combination of nannies, boarding schools, and lessons in self-reliance [10]. For the first two years of Vincent's life, Wu and her husband lived apart during the week, reuniting for weekends. During the week, Vincent lived with his father, under the care of a nanny. Throughout the remainder of their lives, Wu's husband did the majority of the housework [4]. While Vincent later said 'it was an okay way to grow up', it hardly fit Bettelheim's stereotype of the woman who put 'womanly' motherhood ahead of career.

Wu was also an admirer of Lise Meitner, and hosted her on a visit to Columbia University in 1957 [4]. Meitner never married or had children. Cecilia Payne, as an astrophysicist and not a nuclear physicist, was probably not front of mind for Wu, but since she is one of the subjects of this book, her family life should also be mentioned: Payne was married to another astronomer, but was clearly more prominent than her husband. Payne, according to her daughter, while 'uniformly warm' toward her children, was not at ease with babies; her husband did much of the housework [17]. Maria Mayer, the subject of the next chapter in this book and a theorist in nuclear physics, began her scientific career playing second fiddle to her husband, but eventually their roles reversed. When her two children were young, Mayer employed a parenting style more in concord with what Bettelheim had in mind when they were young, stipulating that 'when my children are sick, I am sick' when she was negotiating her contract for a teaching position at Sarah Lawrence College [18]. Nevertheless, the Mayers hired a nanny during this period, since both parents were busy with work related to World War II, and the children felt less cared for as Mayer devoted an increasing amount of time to physics [19]. Each of these female physicists, then, found a different path, in contrast to the single 'feminine' approach that Bettelheim was stressing.

Wu was part of a panel of five female scientists who spoke immediately following Bettelheim. Most of the members of the panel pushed back against Bettelheim in one way or another. Wu went last, and minced no words. After noting that over the preceding hundred years little progress had been made in advancing the rights of women to practice science, she challenged one of the central arguments articulated by Bettelheim:

> Bringing a womanly point of view may be advantageous in some areas of education and social sciences, but not in physical and mathematical sciences where we strive always for objectivity. I wonder whether the tiny atoms and nuclei or the mathematical symbols or the DNA molecules have any preference for either masculine or feminine treatment.

She then went on to note the deep talent pool for science represented by women:

> It was the discovery of radioactivity by Professor and Madame Curie that made people realize the existence of the nucleus. Madame Curie discovered and identified several chemical elements and received not one but two Nobel Prizes, the first time in physics and the second time in chemistry. No man in history has yet equaled that honor and distinction. Her elder daughter, Madame Irène Curie Joliot, and her husband were also awarded a Nobel Prize for their discovery of artificial radioactivity. We are extremely proud of Lise Meitner's achievements. She contributed greatly to our understanding of the alpha and gamma radiations. She worked very closely with Dr Otto Hahn on uranium fission until circumstances forced her to leave Germany. She and her nephew, Dr Frisch, gave the first explanation of what Hahn had observed and named the process of 'nuclear fission', a word borrowed from biology. In 1963 another woman physicist, Dr Maria Mayer, was awarded the Nobel Prize in physics for her important contribution to the nuclear shell model. Never before have so few contributed so much under such trying circumstances! Why should we not encourage more girls to go to science?

Finally, Wu used her biting wit to turn Bettelheim's claim about family responsibilities on its head:

> Professor Bettleheim pointed out that, as much as women want to be good scientists or engineers, they want first and foremost to be womanly companions of men and to be mothers. How can we agree with him any less than wholeheartedly? However, this noble human desire to be devoted companions and good parents must, ideally, be equally shared by men...In our present society of plenty and proficiency, is it too much to provide excellent professional child care during the day so that mothers can get away from monotonous household chores and work in their chosen field [16]?

4.4 Rising through the ranks

After two years as a post-doc at Berkeley, Wu became an assistant professor at Smith College in Massachusetts. She enjoyed teaching at Smith, and continued to attend scientific meetings, but Lawrence wondered whether this was the best use of her talents, asking her if she was happy not being able to do cutting-edge experiments. 'I feel sort of out of the way', she admitted. That was enough for Lawrence; he wrote her letters of recommendation to multiple universities, *all* of which offered her positions, albeit temporary ones [7]. Smith responded by promoting her to associate professor in her second year (a remarkably quick rise), and giving her a large raise [4].

But in addition to her separation from experimental work, Wu felt lonely at Smith. She missed her friends from Berkeley, and saw her husband, who had taken a job at RCA in Princeton, New Jersey, only on weekends [10].

Wu accepted a position from Princeton, although ironically it was to teach, with no research responsibilities [4]. In an interview later in life, Wu claimed to be the first woman to be an instructor at Princeton [4], but that is not strictly true [20]. It's possible, however, that she was the first to teach undergraduates.

Lawrence continued to advocate on her behalf. By this time the Manhattan Project was underway. One branch was indeed located in Manhattan, at Columbia University, in its newly christened Department of War Research. Lawrence recommended Wu for a position there [4].

Interviews for work on the Manhattan Project necessarily involved a bit of the problem of the chicken and the egg. Candidates couldn't be told about the project until they'd agreed to join it, but it was potentially difficult for them to decide whether or not to join until they knew what it was about.

In Wu's case, she was invited to Columbia and questioned extensively about her knowledge of advanced physics. When the moment came for the interviewers to hint at the nature of the project, Wu had already deduced it; her interviewers had not bothered to erase the work on the blackboards in their offices. With that difficulty bypassed, her bemused interviewers asked if she could start the next day [10].

For the remainder of the war, she worked on uranium enrichment and the development of improved gamma-ray detectors [21].

Columbia had a history of being more open to women than Princeton, although only in a relative sense; each new advance for gender equality was a battle, often resulting in compromises that allowed women in, but only in limited or secondary roles. The anthropologist Ruth Benedict was an important trailblazer, becoming the first assistant professor at the University in 1931 and the first to gain tenure in 1937. In physics, the trailblazer was Lucy Hayner, a physicist, who joined the tenure-track in 1946 [22], retiring as a tenured associate professor in 1971 [23].

Thus, at Columbia, Wu was less of a novelty than she had been at Princeton. After serving several years as a research associate, she was hired directly to the rank of associate professor, along with a grant of tenure, in 1952. By 1958, she had achieved the rank of full professor [4].

Nevertheless, Wu's gender still made her very much the exception at Columbia, as at other top research universities. Aside from faculty with joint appointments at Columbia's coordinate women's college, its medical school, or its teacher's college, as of 1960 Columbia had only seven female faculty members in the arts and sciences [24], two of whom were Wu and Hayner.

Was Wu's progress up the academic ladder slowed by her gender? Yes, at times. In 1951, while Wu was still a research associate, one of the other professors in the

Table 4.1. Comparison of career track of Wu and the most famous of her Columbia contemporaries. Includes all Columbia physics faculty in the Array of Contemporary American Physicists [25] with start dates between 1930 and 1959 for whom dates of each stage are listed.

Name	Eventual Nobel?	First Year at Columbia	Pre-Tenure Track (years)	Assistant to Associate Professor (years)	Associate to Full Professor (years)
John Dunning	No	1933	2	3	8
Polykarp Kusch	Yes	1937	9	0	3
Eugene Booth	No	1937	9	1	2
Willis Lamb Jr	Yes	1938	7	2	1
Chien-Shiung Wu	**No**	**1944**	**8**	**0**	**6**
Charles Townes	Yes	1946	0	0	2
Jack Steinberger	Yes	1950	0	2	3
Tsung-Dao Lee	Yes	1953	0	2	1
Melvin Schwartz	Yes	1958	0	2	3

department, Willis Lamb, proposed Wu be made assistant professor. Years later, Lamb said that the only reason the request was denied was that the position of assistant professor required teaching, and Wu was a woman [4]—at the time, Columbia's undergraduate student body was still all male. Still, this objection did not cause excessive delay, as Wu's appointment was approved the next year.

An inspection of table 4.1 shows that Wu's rate of progress was similar to that of the most prestigious of her male colleagues who started at Columbia before the war, but considerably slower than that of those who followed. As a result, she watched Tsung-Dao Lee, who became an assistant professor one year after she was appointed as an associate professor, reach the rank of full professor in 1956, two full years before she did. While not nearly as dramatic as the pauses in the careers of Payne and Mayer, it's possible that she might have been made a full professor a few years sooner had she been male.

While Wu's rate of progress through the academic ranks was slowed modestly because of her gender, the effect on her salary was more dramatic, a discrepancy that was not corrected, despite a succession of department chairs who could have done so, until 1975 [4].

4.5 'Wasting her time'

Students just beginning to learn physics in high school or college are often excited by the idea of constructing theories, or perhaps by experiments that investigate previously unknown phenomena. Introductory physics labs, in contrast, often involve 'verification experiments,' designed to test whether experiment agrees with theory. For these students, this can seem uninteresting—perhaps it needs to be done, but surely, they think, that's not where the action is.

Wu, in contrast, was drawn to such experiments. 'It is the courage to doubt what has long been believed and the incessant search for verification and proof that pushes the wheels of science forward [26]'. When conducted by an expert experimentalist like Wu, verification experiments are often like the period at the end of a sentence, firmly establishing a theory which was previously only provisional. Again and again, Wu conducted experiments of this type, usually finding that the theory being tested was correct and explaining experimental discrepancies found by other laboratories. A sampling of quotations from her many papers:

From a 1947 paper on neutron capture:

> A detailed comparison between the experimentally measured cross section of Cd and the Breit–Wigner one-level formula gives excellent agreement for over a factor of 100 in energy and 1000 in cross section [27].

From a pair of 1949 papers on the beta-ray spectrum of an isotope of copper:

> In view of the large decrease in the deviation resulting from the use of thinner and more uniform sources and the use of a more rigorous Coulomb correction factor, it seems probable that the remaining small observed deviation is instrumental...[28]
>
> The good agreement between the theoretical and experimental curves in figure 2 indicates that the Fermi theory probably does approximate the true distributions for negatrons and positrons at low energies. In any event, the true deviations must be much smaller than has been previously suggested [29].

From a 1950 paper testing a theory regarding gamma rays produced via matter-antimatter annihilation:

> Therefore, it appeared to be highly desirable to reinvestigate this problem by using more efficient detectors and more favorable conditions...the agreement [found by our new, more stringent, experiment] is very satisfactory [30].

Regarding the measurement of x-rays produced during the beta-decay of a promethium isotope in 1954:

> Because of the approximate nature of the theoretical calculations, the agreement between the theoretical and experimental results is considered satisfactory [31].

From a 1956 study of electron capture:

> The agreement [between theory and experiment] is now excellent over the whole energy region [32].

From a 1957 paper on the ground state of atoms of antimatter:

> The experimental values agree with the theoretical value... [33]

From a 1962 paper on electron capture by isotopes of iron and cesium:

> The results are in good agreement with Martin and Glauber's revised calculations including relativistic and screening effects [34].

From measurements of the beta spectra of isotopes of boron and nitrogen in 1963:

> This investigation confirms that the deviations from the allowed shape of the observed beta spectra for B^{12} and N^{12} have the correct magnitude and sign due

to the weak magnetism term. This unique relation between the beta interaction and electrodynamics strongly supports the conserved vector-current theory [35].

Some of these confirmations were routine; others, such as the demonstration of the conservation of vector current in the last excerpt above, resolved multiple conflicts in previous experiments and were treated as major news within the physics community.

In 1956, Wu was faced with the opportunity for another verification experiment.

For several years, particle physicists has faced a mystery, known as the θ–τ puzzle. (See the Science Summary at the end of this chapter for details.) According to the scientific consensus of the time, subatomic particles each possessed a quantity known as *parity*, which could be even or odd, just as they possessed other quantities such as charge and mass. A particle could decay, turning into other particles, but the total mass[3], charge, and parity of the products was always the same as that of the original particle. In other words, quantities like mass, charge, parity were *conserved*.

The recently discovered particles θ and τ decayed into products with a different total parity, and thus were themselves thought to have different parities, making them different particles. But they were otherwise identical—even the average time it took them to decay was identical to within the precision of the experiments that had been conducted.

As physicists focused on the problem, they realized that one possible interpretation was that the θ and τ really were the same particle, but that parity was *not* conserved when they decayed. This was a startling idea, and initially amounted to little more than unfounded speculation [36].

Two scientists, Tsung-Dao Lee of Columbia and Chen Ning Yang of Princeton, decided to pursue the idea. They realized that there was very good evidence that parity was conserved in decays related to what is known as the *strong nuclear force* (the force that holds protons and neutrons together in the nucleus). The decays of the θ and τ, however, are related to the *weak nuclear force*, the same force responsible for beta decay, in which Wu was an expert. Since Lee and Wu were professors in the same department at Columbia in addition to both being immigrants from China, they knew each other well. Lee went to Wu to ask about evidence of parity conservation in beta decay. Wu didn't know the answer immediately, but directed [10] Lee to a recently-released encyclopedic book of nearly a thousand pages, *Beta and Gamma-Ray Spectroscopy* [37], by Kai Siegbahn (whose father, Manne, had treated Meitner so poorly in Sweden).

Lee and Yang carefully went through Siegbahn's book. When they were done they concluded that there was no experimental evidence, implicit or explicit, as to whether parity was conserved in beta decay. The theory of parity conservation wasn't proven wrong by existing data, but it wasn't verified by it either [38]. After consulting with another scientist who suggested that a nuclear polarization experiment might provide a test, Lee went back to Wu, who provided more details of how such an experiment might be executed [9]. In June of 1956, Lee and Yang published a paper indicating what they had discovered thus far, and detailing the experiment

---

[3] In order for mass to be conserved, Einstein's relation $E = mc^2$ must be used to account for any energy released in the decay.

suggested by Wu. In it, they acknowledged Wu, along with three other scientists, for 'interesting discussions and comments' [39].

Should Lee and Yang have invited Wu to be a coauthor on their paper? A tremendous amount of effort on the part of Lee and Yang went into performing the analysis of the data in Siegbahn's book, and in to writing the article itself. Wu, on the other hand, contributed key direction and ideas.

And Lee was one of the professors at Columbia who jumped Wu in academic rank. If she had been a full professor and he an associate professor, perhaps she would have taken on a somewhat larger role in directing her junior colleagues, which then could have been acknowledged by placing her name in the last position in the author list, as is typical in such cases. But with Lee and Yang both being newly minted full professors, that might have seemed less appropriate.

In any case, Wu did not object, either at the time or later, to Lee and Yang writing the paper on their own.

Their paper was submitted in late June. Although it was not published until the beginning of October, a preprint was widely circulated among scientists [40]. Since she was involved in the discussions preceding the paper, Wu had a head-start in thinking about the verification of parity conservation, but not nearly as much as the October publication date would make it appear.

The proposed experiment was challenging, but not impossible. It required a combination of capabilities that did not fit neatly into any single group's research program: it had to be conducted at extremely low temperatures, close to absolute zero, but also required expertise in measuring beta decay. When Lee and Yang's paper came out, only three groups chose to pursue tests of parity non-conservation; Wu's, a group led by Valentine Telegdi at the University of Chicago (Telegdi read a preprint of Lee and Yang's paper in August [40]), and a group of Italian scientists working at Fermi's former institute in Rome.

As with other verifications Wu had performed, she expected the existing theory would be confirmed. In later recollection, she estimated the chance of parity conservation being discarded were 'a million to one' [7]. This seems likely to be a bit of exaggeration on Wu's part—the physicist and future Nobel Laureat Richard Feynman initially considered the odds to be ten thousand to one against, but eventually settled on odds of fifty to one against for the purposes of a friendly wager [4]. Yang himself later said:

At the time I was not betting on parity non-conservation; Lee was not betting on parity non-conservation[.] I don't think anybody was really betting on parity non-conservation. I don't know what Telegdi was thinking, but Miss Wu was thinking was that [sic] even if the result did not give parity non-conservation, it was a good experiment [41].

Telegdi later described it in this way:

It was contrary to everything you had ever learned in quantum mechanics. It was a very bold statement. That's why very few people wanted to do the experiment—because they felt that it was so unlikely. And others thought,

'Well, maybe parity is not conserved, but the effect will be so tiny that it will be too hard to detect' [42].

The theorist Wolfgang Pauli, another Nobel Laureate in physics, minced no words:

A good experimentalist like Wu should find something more important to do, rather than wasting her time in this obvious matter. Everybody knows that parity is conserved [4].

In fact, the Italians were the first to complete an experiment based on the questions raised by Lee and Yang, presenting their results in September. In keeping with the difficulty of the measurement, the results were inconclusive [40].

Despite the likelihood that parity was conserved, Wu placed the experiment on the front burner. She and her husband had a round-the-world trip planned for that summer, including visits to a physics conference in Geneva and their first trip back to China (albeit Taiwan) since Wu had come to Berkeley twenty years before. Wu cancelled her tickets, staying behind to work on the experiment while her husband made the trip alone [26].

Wu needed a collaborator with the expertise and equipment for extremely low temperature work. Richard Garwin seemed the perfect choice: he had gone to graduate school with Lee and Yang, and now worked at IBM's Watson Laboratory, which at that time was sited next door to Columbia. But Garwin was busy with his own projects, and suggested Ernest Ambler of the National Bureau of Standards (NBS) [43].

Ambler agreed to take on the project (figure 4.2), but it soon became evident that he was not used to Wu's hard-driving style of experimentation. While she cancelled her planned vacation, he took two weeks off in August. Wu's graduate students, too, did not get along well with Ambler's team, with the result that Wu and Ambler agreed to have NBS personnel fill out the team, with Wu as the only representative from Columbia.

The experiment was complicated, and required considerable preparation. June and July were spent designing detectors that would be accurate in ultra-low temperatures and high magnetic fields [26]. August, while Ambler was away, saw Wu and her team at Columbia computing the effects of magnetic fields and scattering on the counts.

Once the experiment was finally underway, they found that the sample was not staying cold for long enough [26]. They decided the problem could be addressed by enclosing the sample within a larger crystal to protect it from heat [44]. They would need to grow the large crystals themselves and, if it were not to result in a long delay to the experiment, do it in record time [45]. After consulting with crystal chemists and being dissatisfied with what they said was possible, Wu and Marion Biavati, her graduate student, borrowed a decades-old book from the chemistry department to teach themselves how to grow crystals, and then developed a new technique for growing excellent, large crystals rapidly [26]. Ambler marveled at their results, describing them as 'crystals as beautiful as diamonds' [4].

Figure 4.2. Wu and Ambler. Reprinted courtesy of the National Institute of Standards and Technology, U.S. Department of Commerce.

Next, Wu and her team had to take two large crystals, and drill a hole in each so that when placed face to face a sample could sit in the space formed by the alignment of the two holes, and then seal the two halves to each other. The sample would thus be entirely cut off from the external environment. Drilling holes in crystals without shattering them is challenging, but they learned from a specialist in crystals that a dental drill would work [26]. By the time they were ready to try again, it was November. At first, they thought they had a result, but it was a false alarm—the adhesive they were using to glue the two halves to each other failed at low temperature, causing the assembly to collapse. The team adapted by using nylon wires to sew the holes shut [4]!

Finally, in mid-December, they began to get real results: parity conservation was violated in beta decay.

Fifty to one, ten thousand to one, a million to one, a waste of time—whatever the odds, it was an improbable result, the kind of thing that a brilliant scientist might come across once in her career, if she were lucky.

Wu needed to double-check. Reporting such a result brings instant fame; having to retract it due to some mistake in the experiment lasting embarrassment. Checking and rechecking, working their way through an agreed upon set of tests through all hours of the day and night, the result became official at 2 am on January 9. The NBS team broke out the champagne—literally, drinking it from paper cups. Whether Wu, usually a teetotaler, joined them in their celebratory drink varies between accounts [4, 26, 44].

The next task was to write the paper, and quickly, as word of the results was spreading to other groups that could extend the experiments. The exhausted NBS scientists waited until Sunday, January 13 to discuss the writing of the paper with Wu. By that time, she had already written the entirety of it herself. All that was left was to choose the order of names. Alphabetically, with Ambler first and Wu last? McGrayne's biography of Wu captures the decision:

With a sigh, Wu indicated that [alphabetically] would not be the correct approach. So, 'like the perfect bloody Englishman', Ambler asked, 'Would you like to go first, Miss Wu?' [10].

Years later, Ambler stood by the decision to list Wu first, as she was both the senior scientist on the team and the one who had initiated the idea of the experiment [10].

While Wu's aggressive style of leadership damaged her relationship with the NBS team, she needed to push them that hard to ensure priority for their discovery. On January 4, future Nobel Laureate Leon Lederman learned of their preliminary result, and realized he might be able to use Columbia's cyclotron to do a much quicker experiment to test for parity violation. He called up Garwin—the same Garwin who had turned down Wu months ealier—and proposed the experiment, which they accomplished over the next four days, finishing on January 8 [46]. Garwin's initial thought was that he and Lederman has just won a Nobel Prize.

At that moment, Garwin and Lederman made an honorable and mature decision. As Garwin put it, 'we were actually ready with our paper before they were, but we would never have done the experiment had she not had her results...she deserves the priority'. He then added another practical consideration: 'Besides, T D Lee would have killed us' [43]. The two groups coordinated so that their papers would both be published as received on January 15. In addition, Garwin and Lederman included this acknowledgment in their paper, along with a citation to the paper of Wu and her NBS collaborators:

We are also indebted to Professor C S Wu for reports of her preliminary results in the Co^{60} experiment which played a crucial part in the Columbia discussions immediately proceeding this experiment [47].

This made it clear that priority of the discovery of parity non-conservation belong to Wu's group and Wu's group alone.

There remained, however, Telegdi's group at the University of Chicago. Their experiment was not complete, but they rushed to submit what they had. Their paper was received two days after the pair of papers from Columbia [48]. Telegdi had lost the race.

Telegdi was not gracious.

Just before Wu wrapped up her experiment, Telegdi's work was interrupted by personal matters. He blames his loss on this delay:

Probably if I hadn't gone to Italy, where my father lived and where he died, we would have finished the real work, let's say, two weeks before, at most. And then we would have gotten in print either earlier or at the same time as our competitor, Miss Wu [42].

This, however, is unlikely. As Telegdi himself indicates, his colleague Jerome Friedman continued work on the experiment in his absence, so the delay was not as serious as Telegdi makes it out. And crucially, Telegdi rushed his work *in order to beat Wu*. He didn't really lose out by just two days, because at the time he submitted his work, it was incomplete. In fact, the paper, as submitted, was underwhelming. According to the editor of *Physical Review*, the journal in which all three papers appeared:

The obstacle encountered by Telegdi was of a different nature. A careful study of his Letter shows that at the time he submitted it the work was unfinished. He obtained an asymmetry of 0.062 ± 0.027 with 1300 events, that is $(690 - 610)/1300$ with a standard deviation of 35 in the difference. The effect is only a little larger than two standard deviations. This should be compared with the overwhelming and compelling evidence presented in the two other Letters as seen especially clearly in their graphs. An effect of less than three standard deviations is quite insufficient in such an important and subtle experiment... Several weeks after submission, Dr Telegdi added a note in proof to his letter, stating that a total of 2000 events had now given an asymmetry of 0.091 ± 0.022, that is $(1091 - 909)/2000$. This is quite an improvement. Note, however, that the 700 additional cases must have shown a surprisingly large asymmetry, namely $(401 - 299)/700$ of 0.146 ± 0.039, as compared to the original 0.062 ± 0.027. These large statistical errors (27%–43%) definitely prove the preliminary nature of the initial result submitted for publication [49].

Since Telegdi did not get his improved data in for 'several weeks', he was actually behind by at least that much.

But we can go further. According to an analysis by the physicist Allan Franklin, Wu's results corresponded to 13 standard deviations, and Garwin and Lederman's to 22, while even Telegdi's revised data only amounted to 4 [50]. The 2.3 standard deviations of Telegdi's initial data would have occurred due to random fluctuations more than 2% of the time, even if parity were in actuality conserved. Feynman's 50 to 1 odds would have been foolish to pay off his bet if Telegdi's initial experiment had been the only evidence of parity non-conservation, as there would be more than 1 chance in 50 of getting a false positive!

In terms of timing, it was the Italian group that beat everyone else to the punch, not just by a few weeks, but by several months. But their result was only 1.3 standard deviations, a result that would have occurred by chance almost 20% of the time, even if parity had been conserved. For that reason, the Italian results didn't attract much notice at the time [40].

Tegeldi's revised results are a bit more convincing, with 0.004% of them occurring by chance, or 1 chance in 25 000. But that's still close enough to Feynman's original ten thousand to one estimate to at least make it an argument whether the error is more likely to lie with the experiment or the theory. And that's assuming the experiment was *perfect*, with no systematic biases. Extraordinary claims require extraordinary evidence, and Telegdi's revised results were still not extraordinary.

The odds of Wu's results occurring by chance, however, would be just a bit more than a million billion billion to 1. It was even enough to convince Pauli [4]!

Telegdi, though, was bitter. He felt the *Physical Review* editors were under the influence of 'non-scientific factors', and that the 'Columbia University gang' had played a trick on him [4]. In a fit of anger, he resigned from the American Physical Society, the publisher of the journal [50], and threatened to leave physics altogether [4].

It is certainly true that Wu had a head-start of Telegdi, because she knew about Lee and Yang's results first. This is hardly evidence of unethical collusion, however, as she contributed to the ideas that went in to the paper! The rest of Telegdi's argument is sour grapes.

4.6 Instant Nobel

Most of the time, there is a considerable gap between when a new theory is proposed and when a Nobel is awarded for it, in part because the Nobel committee does not want to award a prize for a theory that is quickly overturned. Fermi's award for discovering 'new transuranic elements' was a rare mistake of this kind, caused by thinking that the discovery was purely experimental in nature, without considering the theoretical framework that was used in its interpretation.

But in this case, the process proceeded with lightning speed. Nominations for each year's prize must be postmarked by February 1. The papers demonstrating parity violation did not actually appear in print until February 15. But the Columbia physics department held a press conference on January 15 [51], which resulted in a front-page story in the New York Times the next day [52]. There was just enough time for Lee and Yang to receive nominations. It is possible that some nomination ballots included Wu's name as well, but that's still unknown; while Nobel nominations are generally made public fifty years afterward, they are not released if the nominees are still alive, which is the case for both Lee and Yang as of this writing.

On October 31, 1957, the Nobel Prize in physics was awarded to Lee and Yang [53].

There is no question that Lee and Yang deserved the Nobel; their work was thorough and provocative, and led directly to experiments which overturned what had been seen as a fundamental law of physics. But that experiment was performed by Wu, and Nobel Prizes allow up to three recipients. Without experimental confirmation, Lee and Yang would surely not have received the prize for their controversial proposal. Why did Wu not share in the prize?

Since the nomination archives and reports are not yet available, we can only speculate.

With Lee and Yang taking two spots, that left only one open. Therefore, it is worth considering who else might have been in competition for that third spot. Too many strong competitors might lead the Nobel committee to decide that it would be better to limit the award to the two theorists, rather than pick a favorite among a crowded field.

Some have thought of Ambler, Wu's collaborator at the NBS. But despite ill-feelings between Wu and the NBS scientists during and after the experiment, Ambler repeatedly and publicly deferred to Wu as the leader of the project, and her name appeared first on the paper announcing their results. It is clear that Wu took precedence over Ambler.

Garwin and Lederman also played a role, but only after they became aware of Wu's preliminary results. They also deferred to Wu in the paper they published. It is unlikely that nominators saw Garwin and Lederman as competition for Wu's potential spot.

That leaves Telegdi and his collaborator Friedman. (Friedman, while he did eventually win a Nobel for a different project in 1990 [54], was very junior at the time, having just completed his doctorate.)

Telegdi was making a lot of noise, fighting with the editors of *Physical Review* and then resigning from the APS altogether, while simultaneously promoting his own case energetically. In late January, for instance, Telegdi sent Pauli preprints of all three experimental papers; by doing so, he was framing the experimental discovery as simultaneous and, at least in his case, independent. This was the first Pauli had heard of the results. Telegdi also appeared as a featured speaker at the special session on parity non-conservation held at the New York meeting of the APS, along with Wu, Lederman, and Yang [51]. I'm not sure who chose the speakers for that session, but again the message was clear: Telegdi was to be treated as having independently and simultaneously demonstrated parity non-conservation.

One relatively recent interview with Telegdi [55], conducted not long before his death, demonstrates both his falsehoods and his venom. It is worth examining some of its claims.

The interviewer broached the topic of parity non-conservation with a mention of Wu:

Lee and Yang came up with the idea that parity might be violated in the weak force. They also suggested experiments, for example, the one that Madame Wu did with beta decay…

Telegdi interrupted angrily:

You know, nothing upsets me more than when you or anyone else refers to Madame Wu in this way because I can tell you the whole story.

Telegdi then claimed that the experiment his group did was 'embarrassingly simple and anybody could have done it'. (This, of course, raises the question as to why the Italian group failed to get significant results, and his own efforts initially only produced results of marginal significance.) He followed by saying that:

> You and most of mankind refer to [the experiment that Wu devised and led] as Wu's experiment. Well, that is very romantic but it is false. You see, in order to do this experiment, you have to align nuclei, which in 1956 was an art, a technique known only to a handful of people in the whole world. In fact, hardly to anybody in the United States. The two people at the Bureau of Standards, one of whom was Ambler, had been imported to the Bureau of Standards from Oxford, because they had this monopoly in their hands. So Madame Wu had to look for somebody who knew how to align nuclei...So she proposed to these people that they do their experiment together. The heavy part, the significant part, the difficult part was done by these people at the Bureau and not by her. Her specialty was radioactivity, she knew how to count the beta rays that would come out; but about the alignment technique that was the crucial part of the experiment, she knew strictly zero. So to give full credit to her is a crime!

This description is so distorted as to constitute a falsehood. Her role in the experiment was not just to count the beta rays that come out! As described in section 4.5, this experiment was more complex than those that Ambler and his colleagues typically performed, and Wu and her graduate student contributed crucially to creating the conditions where the nuclei would stay aligned long enough for the experiment. It's true that she knew little about aligning nuclei when she began the work, which makes her accomplishments all the more impressive.

In addition, the radiation detectors had to be designed so that they could work at very low temperatures; detector design was another specialty of Wu's [26].

It is also true that Wu would have had a difficult time completing the experiment without Ambler and his team, but for Ambler and his team to complete it in a reasonable period of time without Wu would have been nearly impossible.

Is it a crime to describe it as 'Wu's experiment'? If so, then it is certainly a crime to refer to 'Rutherford's gold foil experiment', which was as much due to Hans Geiger and Ernest Marsden as to Rutherford. And to think that 'Werner von Braun's rockets' were the effort of von Braun alone is laughable. Wu was the leader of the experiment and a crucial contributor; there should be nothing controversial about describing the experiment as hers.

In the same experiment, Telegdi claims that 'hardly anyone in the United States' knew how to align nuclei, and that the NBS team had a 'monopoly' on this ability. The fact that the NBS team was Wu's *second* choice, after Garwin, suggests this is an exaggeration.

Telegdi continued with a claim that Lee told Wu 'what nucleus would be particularly suitable', a bizarre inversion of the truth. He said that Wu 'was hardly ever in Washington; she would come every now and then for a few days but to call it the "Wu experiment" is criminal'.

In fact, Wu travelled every week between Columbia and Washington, and was there continuously during the crucial checks in late December and early January [4]. Telegdi next addressed the author list:

> At least give all the members of the group equal credit and do that in alphabetical order…she did the horrible thing; she put her name first, although it starts with a W. The others were English gentlemen and did not have the courage to object.

Although alphabetical order for authors is standard in some fields, that is not the case in nuclear physics. Telegdi twists Ambler's wry description of his acquiescence to Wu being first in to a kind of failing on the part of Ambler. It is worth noting that while Telegdi himself usually used alphabetical order, in 1961 he coauthored a paper with Roland Winston, and Winston's name came first, even though it started with a W [56]! Presumably Telegdi did not have the courage to object.

But, if these experiments were so easy, why didn't Telegdi simply finish first? His answer:

> It was just at the time when my father died and I had to come and I left Friedman alone. This created a certain amount of problems. A similar experiment was done at Columbia University with electronic techniques, by Garwin and Lederman. Because of my travelling we got to the journal with our paper a day or two later than they did.

One is tempted to feel some sympathy for the grieving Telegdi, taking time to visit his grieving mother at a crucial junction. But this is another misrepresentation. Telegdi's father died on September 3; he did not leave for Italy until December, when the semester ended [57]. Apparently Telegdi could not interrupt his courses or other duties to tend to family matters, but could delay his experiment [55][4]. Portraying this as a choice between familial responsibilities and the experiment is simply false. It's true that Wu prioritized her experiment more highly than Telegdi did his…and thus it is not surprising that she finished first.

In addition, suppose that Telegdi had not been travelling? When he heard rumors of results coming from Columbia, he might have rushed his paper even more, and would not have trailed the other two papers in received by date. But if that had been the case, his experiment would have been even more incomplete, and would have looked even more like the earlier, Italian result.

---

[4] Strangely, Telegdi's coauthor Friedman gives a startlingly different description of events than Telegdi himself did, saying that Telegdi 'had to go to Europe during the early autumn of 1956 on personal matters and remained there for about two months'. Telegdi writes that he 'could not go to Europe before the Christmas vacation'. The stories of these two scientists centrally involved in the same experiment at the same site are simply irreconcilable. I have chosen to take Telegdi's description as accurate, because he presumably remembers not being able to visit his grieving mother for three months; also, Telegdi put his account in print while Friedman's was in an oral history. Perhaps Friedman is conflating events that occurred with different experiments.

Telegdi also developed a life-long feud with Lederman, and so took pains to distort his contributions as well, crediting 'all the key ideas' to Garwin, and injecting a note of ethnic pride by noting that Garwin, like Telegdi, was of Hungarian descent. He then scrambles the roles of everyone involved. I've put the correct version in {braces}:

> For a long time Lee {actually Wu} tried to persuade Lederman {actually Garwin} to do this experiment and Lederman never knew how to do it. Then Dick Garwin {actually Lederman} heard about this during a lunch conversation and told them how to do it and they did it in one night {actually four days and nights}...It is also said that they wanted to get into print before Madame Wu but they were stopped from doing that. {they stopped themselves}

Telegdi was not done with his excuses:

> But the key thing which we do not read very much about in the literature, is that when Garwin and Lederman did the experiment, they already knew a preliminary result of the cobalt experiment in Washington. They knew that parity was violated in cobalt decay—which I did not know.

For once, Telegdi is presenting an accurate version of events. But in his eagerness to make excuses, he strengthens Wu's claim to priority. The implication is that if he had known about Wu's preliminary results, it would have helped him move faster. He did know by the time he sent in his paper, because the press conference at Columbia had been two days before. And the results he then sent in were themselves preliminary, because he added to them in the interval between submitting his paper and its publication. And gathering that additional data took weeks, not the four days taken by Garwin and Lederman, and even then was not as definitive.

There is no ambiguity, and should be no historical controversy. Wu and her colleagues at the NBS were clearly the first with a definitive experiment demonstration of parity non-conservation.

But sometimes the squeaky wheel gets the grease. Telegdi's furious pushback resulted in his revised paper being published with a footnote:

> For technical reasons, this Letter could not be published in the same issue as that of Garwin, Lederman, and Weinrich [48].

'Technical reasons' sounds like an error on the part of the journal, when in reality it meant that their experiment was not sufficiently complete for publication.

When there are fierce disputes in scientific priority, often the resolution is to consider all claimants to have made the discovery independently; Leibniz and Newton, for instance, are both credited with the invention of calculus. In many accounts, this has been done with parity non-conservation. Franklin's otherwise excellent account of the experiments, for example, ends with this baffling statement:

All of the histories of this episode, however, including this one, grant equal credit to the three experiments [50].

For the record, the history that you are currently holding in your hands does not. Another source which unambiguously gives credit to Wu is Yang's Nobel lecture. While Lee was a faculty member at Columbia, and thus Wu's colleague, Yang was at Princeton, with a more attenuated relationship to her. Yang, Garwin, and Telegdi's coauthor Friedman had all been graduate students at Chicago, where Telegdi worked, so it's unlikely that Yang would downplay their contributions had they been crucial. And yet they are not mentioned in his Nobel speech. Instead, he says:

> This experiment was first performed in the latter half of 1956 and finished early this year by Wu, Ambler, Hayward, Hoppes, and Hudson. The actual experimental setup was very involved, because to eliminate disturbing outside influences the experiment had to be done at very low temperatures. The technique of combining β-decay measurement with low temperature apparatus was unknown before and constituted a major difficulty which was successfully solved by these authors. To their courage and their skill, physicists owe the exciting and clarifying developments concerning parity conservation in the past year [38].

During the two weeks in January between the Columbia press conference and the due date for Nobel nominations, Telegdi had muddied the waters. If nominators, and after them the Nobel Committee for Physics, accepted Telegdi's attempts to gain equal status with the Columbia teams at face value, then there were too many experimentalists to have them share Lee and Yang's prize. Even if they realized that there were conflicting narratives, how were they to sort it out so quickly? Safer to just limit the prize to the two theorists.

Probably some nominators did list Wu along with Lee and Yang, but we won't know the full pattern of nominations until the last of the trio passes away.

After the prize was awarded to the two theorists, Telegdi sometimes employed a different argument against Wu:

> If an experimentalist performs an experiment with known techniques and on top of that the experiment has been clearly suggested by the theorists, where is the merit? This is true for me, too. It could easily have been the case, with slightly different circumstances, that we would have gotten the result first. If we could have done our emulsion work a little quicker, if we had done [sic] a few more scanners, etc., even then it would not have contributed to our cleverness, or glory. None of the experimentalists deserved the Prize in this case [55].

The argument Telegdi employs here could be used for any case where there is a race between experimental groups to be the first to obtain a result suggested by theorists. But the Nobel Prize *is* sometimes given to the winners of that kind of race. In 1959, for example, the prize was awarded to Segrè and Owen Chamberlain for being the winners of the race to produce antiprotons [13].

I have been hard on Telegdi, and devoted considerable space to demonstrating his apparent malice and falsehoods. That is because most accounts of the parity experiments accept one or the other of his arguments, e.g. that it was a virtual three-way tie, or that his experiment was delayed by his father's death, or that the experiments were straightforward, and I would like to set the record straight, or at least invite readers to consider the evidence for themselves.

But before I leave the topic, there is one more piece of information to provide. This does not bear directly on Telegdi's arguments, or on Wu's merits. But it does, I think, say something about Telegdi.

His experiment consisted of carefully examining the tracks left by particles as they traversed a photographic emulsion. This was very time-consuming, particularly since he hoped to measure 2000 events. It would indeed have been a heroic effort if he and Friedman had counted all those events themselves.

But they didn't...in fact, the experiment wouldn't work if they did. That's because in this kind of work people who know the result they want will tend to subconsciously bias their measurements.

So they used a 'modest group of scanners' to do the job [57]. The scanners were not electronic but, like the computers at Harvard a half-century before, people, led in this case by Elaine Garwin [58], sister-in-law of Richard Garwin of Columbia's 'weekend experiment'. Telegdi was relying on them to perform the actual measurements, just as Pickering had relied on his computers. But unlike Pickering (at least in his later years), Telegdi barely acknowledges their contributions, except to say he would have worked more quickly if he had more of them. When he speaks of the work going more slowly when he was in Italy, it was not because he was not available to perform scans, because he wasn't available to perform scans anyway. The most likely explanation is that he felt that if he were present he could have pushed his scanners to work faster, or work longer hours, or that he could have obtained more of them. While Wu also drove the team at NBS very hard, they at least were senior scientists who became coauthors on the paper, and Wu shared in the hard work and long hours involved in the measurement.

Chemist and historian of science Magdolna Hargittai lists another reason why Wu may have been denied the prize: timing. Nobel's will specifies:

prizes to those who, during the preceding year, shall have conferred the greatest benefit to mankind...one part to the person who shall have made the most important discovery or invention within the field of physics... [59]

The timing provision has been interpreted very liberally; in practice, prizes are often given years or decades after the relevant discovery or invention, perhaps using the argument that it is the year that the discovery or invention becomes important that matters [60].

Hargittai writes that 'the rules stipulate that the awarded work must have been published *before* the year of the prize', and that therefore Wu was ineligible for the 1957 prize, because her paper was published in January of the same year [19]. Nobel's will, the guiding document for the allocation of prizes, makes no mention of publication at all. Turning this around, why were Lee and Yang eligible in 1957? The importance of their work was not clear to anyone until Wu and the others got their results. If Wu was not eligible because of timing, it is a difficult needle to thread to argue that Lee and Yang were.

Wu had another good opportunity to win the prize for the parity experiment she led. In 1980 James Cronin and Val Fitch won the Nobel Prize in physics for a 1964 experiment in which they showed that the product of charge and parity is also violated [61]. It would have been reasonable for them to split the prize with Wu for her 1956–57 experiment at that time. That the 1980 prize was given to Cronin and Fitch alone suggests that it was arguments of the kind Telegdi raised, rather than those of timing, that were the primary obstacle for Wu.

4.7 Honors

In 1973, Wu became vice-president elect of the American Physical Society (APS), a position which led directly to her becoming president of the APS in 1975. All the previous presidents had been white males, meaning that she broke two barriers at once. By this time, Wu was very outspoken about the difficulties faced by female physicists in America, and speculated often about the reasons that women made up such a small proportion of American physicists, including poor guidance counselors [7], lack of child care, and the expectation of many men that their wives would bear the brunt of the housework [4].

While she never received a Nobel, Wu did receive many other honors, including the first ever Wolf Prize in physics (now widely considered the second most prestigious physics prize after the Nobel), the National Medal of Science (bestowed on her by President Ford), and the Comstock Prize in Physics, which is awarded only once every five years [4].

4.8 Science summary: parity

From at least the time of Galileo, physicists have relied on symmetries to help make sense of their understanding of nature. To a physicist, a symmetry is any operation that makes no difference to the system being studied. For example, suppose we have an elaborate system of pulleys and weights. We measure its behavior on a Tuesday at 8 am. We measure its behavior again on Thursday at 2 pm, and on Monday at midnight. If it always behaves the same, we say it is symmetric under time translation.

Physicists have long believed that the laws of physics themselves are symmetric under time translation. Thus, if the pulley system behaves differently at midnight on Monday than it did on Thursday at 2 pm, it must be due to something changing in the system or the environment; perhaps the temperature of the room was higher on Thursday, or perhaps the pulleys have rusted in the interim. But if we could re-create the same system and the same environment, we should, according to the prevailing paradigm in physics, get the same results. This is necessarily a somewhat abstract idea; physical laws are never measured directly, but by examining the behavior of systems placed in some environment. Real systems and environments tend to change over time. But the accumulated experience of millions of experiments suggests that the laws that govern their behavior do not change.

In 1915, Emmy Noether, a mathematician at the University of Göttingen, proved that for every symmetry there is a corresponding quantity which is conserved (i.e. does not change) [62]. For time translation, that quantity is energy. Saying that the laws of physics do not change over time is entirely equivalent to the statement that the energy of the universe is conserved.

Similarly, the laws of physics do not appear to change from place to place, as long as the environment is contrived to be the same. The gravitational field, for instance, is different on Mars than on the Earth, but that is because of differences in the environment caused by the two planets. If the gravitational field and other environmental effects on an experiment are contrived to be the same, then the laws of physics will turn out to be the same. This symmetry—translation in space—corresponds to the conservation of momentum.

Noether's theorem provided a powerful shortcut for discovering new conservation laws. If a symmetry could be found, then Noether's theorem led straight to the corresponding conservation law.

When non-scientists think of symmetry, they do not usually think of symmetry under translation in space or time. Most commonly, they think of left–right symmetry, like an ink-blot (figure 4.3). People say ink-blots are 'symmetric', but in a physicist's language the statement should be more specific: they are symmetric under reflection.

If we look in a mirror, we can imagine a whole 'looking-glass' world, which is the mirror-image of ours. If we took a system and environment from our world, and put it into the looking-glass world, would it behave the same? In other words, would the laws of physics still be the same?

Overwhelmingly, the intuition of physicists told them that it would. Yes, there are many particular systems that are not symmetric with respect to reflection. If a looking-glass surgeon were to try to perform surgery on one of us, she would be surprised to find our heart on the 'wrong' side. Biological molecules such as DNA also lack reflection symmetry. But that could be an accident of evolution; people with hearts on the left side of their chests have children with hearts on the left side of their chests, so once the population is established, perhaps by chance in some ancestral organism long-ago, the pattern will persist. But it seems reasonable to think that a mirror-image person, eating mirror-image food, would be able to function just fine under *our* physical laws. Our *food* might be poison to them, because many of the molecules that make up food are not symmetric under reflection, but our *physics* would be the same.

Since physics was thought to be symmetric under reflection, there must be a corresponding quantity which is conserved. That quantity is called *parity*.

Figure 4.3. This image is symmetric with respect to reflection across the dashed line.

There is an important way in which reflection symmetry is different from symmetry under time or space translation: two sequential reflections always leave us back where we started. This in turn means that while energy and momentum can take on any values, there are only two distinguishable states of parity. These can be labeled positive and negative, or even and odd. In the rest of this section, we'll use even and odd.

The terms even and odd are a kind of mnemonic that tells us how parities combine. In arithmetic, adding two numbers that are both even (e.g. $6 + 8 = 14$) or both odd (e.g. $5 + 11 = 16$) yields an even number; adding an even number to an odd number (e.g. $6 + 11 = 17$) yields an odd number. Similarly, the total parity of two particles is even if the individual particles either both have even or both have odd parity, while the total parity is odd if one of the particles has even parity and the other odd.

In the early fifties, many new particles were being discovered. One, the τ meson, was known to decay into three pions. From other experiments, pions were known to have odd parity. Just as adding three odd numbers yields an odd number, the total parity of three pions is odd. As long as parity is conserved in the decay of the τ meson, it must have odd parity too.

Another newly discovered particle, the θ meson, consistently decays into two pions. Since pions have odd parity, the combination of two pions would have even parity, suggesting the parity of the θ particle was even.

As the properties of the τ and θ mesons were investigated, a curious thing was discovered: other than the fact that one decayed into three pions and the other into two, they appeared identical, with the same mass and the same lifetime.

Particles were already known that sometimes decayed into one set of products and sometimes another; nothing in the laws of physics prevented that. But in each of the decays quantities like the total energy and the total momentum were always conserved. The τ and the θ were decaying into products with different total parities. Either the τ and the θ were different particles, or, if it was one particle with two different options for decay, parity was not conserved for one of the decay paths.

If parity is not conserved, that means the laws of physics are not symmetric under reflection.

For readers who have taken some physics in high school or college, it's worth expanding on what that would look like. After all, high school and college physics courses seem to suggest that the laws of physics can tell left from right, because to apply those laws students are required to learn a series of 'right hand rules'. If you use your left hand on an exam by mistake, you get the questions wrong!

Let's examine one of those rules from introductory physics in more detail.

Suppose there is an electrical current consisting of positive charges travelling in a narrow beam from the floor up toward the ceiling. According to what is taught in introductory classes, this current creates a magnetic field in a circular pattern around the beam. This circular pattern can be seen by, for example, the use of iron filings (figure 4.4).

The iron filings are aligning with the magnetic field, which circles the current. But does it circle clockwise or counterclockwise? To decide, students are taught to use the right-hand rule shown in figure 4.5.

But this is an arbitrary definition; the iron filings don't show direction.

The direction of the magnetic *field* can then be used to find the magnetic *force* on, for example, a stray positive charge travelling upward, parallel to the main beam. To find the direction of the magnetic force, a second right-hand rule must be used (figure 4.6). The result of applying the second right-hand rule to the first is that the magnetic force on the stray charge is back toward the main beam.

The direction of the magnetic force can be measured directly; the direction of the magnetic field cannot.

In fact, suppose both right hand rules were replaced with left hand rules. If that were the case, we would still find that the stray charge felt a magnetic force pointing back toward the main beam; the physical result would be exactly the same. Thus, the laws of magnetism, as far as they can be measured, *are* symmetric under reflection, even though the way we formulate them in introductory physics courses does not at first appear to be.

As far as we know, all the laws of electromagnetism are symmetric under reflection.

Lee and Yang, however, were speculating about a different fundamental force: the weak nuclear force, which is responsible for the beta decay that some radioactive nuclei undergo.

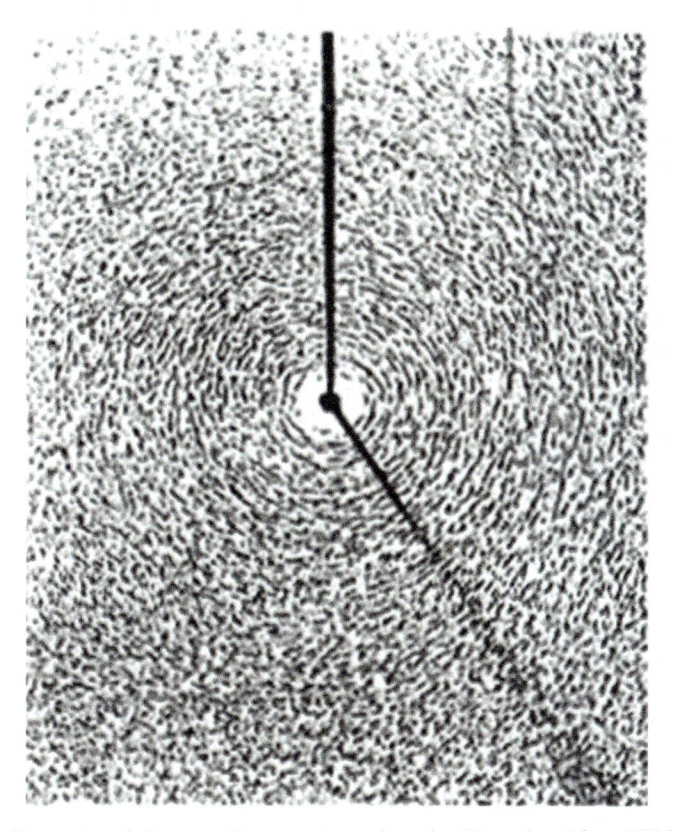

Figure 4.4. Iron filings trace circles around a current-carrying wire. Reproduced from [63] by permission of Taylor and Francis Group, LLC, a division of Informa plc.

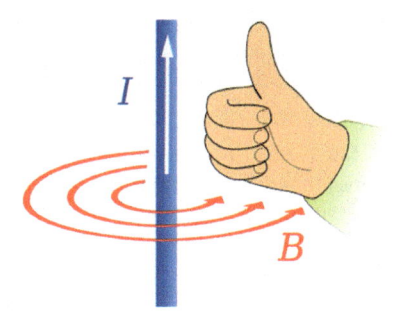

Figure 4.5. Right-hand rule for direction of magnetic field B around current I. Figure from Jfmelero at https://commons.wikimedia.org/wiki/File:Manoderecha.svg, available under a Creative Commons Attribution-Share Alike license (https://creativecommons.org/licenses/by-sa/4.0/).

Wu and her colleagues conducted an experiment using Co^{60} nuclei, which undergo beta decay. Using an external magnetic field, the Co^{60} nuclei can be lined up so that they are mostly spinning the same way. Random thermal motion will tend to jumble up the direction of spin, so the nuclei have to be kept very, very cold. When the Co^{60} atoms decay, they eject an electron. If we imagine the electron, as it

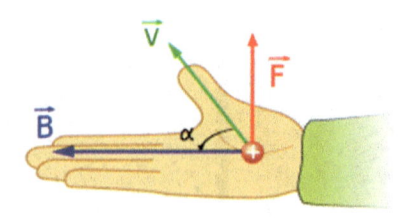

Figure 4.6. Right-hand rule for direction of magnetic force F on a positive charge moving with velocity v in magnetic field B. Figure from Jfmelero at https://commons.wikimedia.org/wiki/File:Regla_mano_derecha_Lorentz.svg, available under a Creative Commons Attribution-Share Alike license (https://creativecommons.org/licenses/by-sa/3.0/deed.en).

flies away, looking back at the spinning nuclei, it would see the nuclei either spinning clockwise or counterclockwise depending on which direction it happened to leave in. (As an analogy, the Earth spins counterclockwise as from a rocket leaving Earth from the north pole, but clockwise as viewed from one leaving from the south pole.) If more of the electrons saw one direction of spin than the other, that would mean something in the mechanism depended on the difference between left and right, and that fundamental physical processes such as beta decay were not symmetric under reflection.

In Wu's experiment, almost all of the electrons left in the direction where they would 'see' a clockwise spin. This means the laws of physics are not the same under reflection, that the τ and θ mesons are just the same particle (now known as the K^+) decaying in different ways, and that parity is not conserved.

References

[1] Wang Z 2007 Wu Chien-Shiung *New Dictionary of Scientific Biography* ed Koertge Noretta (New York: Charles Scribner's Sons)

[2] Segrè E and Wu C S 1940 Some fission products of uranium *Phys. Rev* **57** 552

[3] Wu C S 1940 Identification of two radioactive xenons from uranium fission *Phys. Rev.* **58** 926

[4] Chiang Tsai-Chien 2014 *Madame Wu Chien-Shiung: The First Lady of Physics Research* (trans) Wong Tang-Fong (Hackensack, NJ: World Scientific)

[5] Kałużyńska I 2014 Male names of women and female names of men in the Chinese society *Els noms en la vida quotidiana. Actes del XXIV Congrés Internacional d'COS sobre Ciènes Onomàstiques/ Names in daily life. Proceedings of the XXIV ICOS International Congress of Onomastic Sciences* ; Joan Tort Donada and Montserrat Montagut i Montagut 791–7

[6] Woo Louie E 1998 *Chinese American Names: Tradition and Transition* (Jefferson, NC: McFarland)

[7] Gilbert L and Moore G 1981 *Particular Passions: Talks with Women Who Have Shaped Our Times* (New York: Clarkson N Potter)

[8] Wong Yin Lee 1995 Women's education in traditional and modern China *Women's History Review* **4** 345–67

[9] Benczer-Koller N *Chien-Shiung Wu, 1912—1997, Biographical Memoirs of the National Academy of Sciences*

[10] Bertsch McGrayne S 1998 *Nobel Prize Women in Science: Their Lives, Struggles, and Momentous Discoveries* 2nd edn (Washington: Joseph Henry)

[11] League Docent Tour Video *Student Life: University Unions: University of Michigan*, accessed April 9, 2017 https://uunions.umich.edu/article/league-docent-tour-video
[12] Michigan Union *The University of Michigan Millenium Project*, accessed April 9 2017 http://umhistory.dc.umich.edu/mort/central/west%20of%20state/Michigan%20Union/index.html
[13] Segrè E 1993 *A Mind Always in Motion: The Autobiography of Emilio Segrè* (Berkeley, CA: University of California Press)
[14] Heilbron J L and Seidel R W 1989 *Lawrence and His Laboratory: A History of the Lawrence Berkeley Laboratory Vol. 1* (Berkeley, CA: University of California Press)
[15] Gottlieb R 2003 The Strange Case of Dr B *The New York Review of Books* February 27, 2003 http://www.nybooks.com/articles/2003/02/27/the-strange-case-of-dr-b
[16] Mattfeld J A and Van Aken C G (ed) 1965 *Women and the Scientific Professions: The MIT Symposium on American Women in Science and Enginnering* (Cambridge, MA: MIT Press)
[17] Haramundanis K 1996 A personal recollection *Cecilia Payne-Gaposchkin: an Autobiography and Other Recollections* 2nd edn edited by Katherine Haramundanis (Cambridge: Cambridge University Press) 39–69
[18] Mayer P C 2011 *Son of (Entropy)^2* (Bloomington, IN: AuthorHouse)
[19] Hargittai M 2015 *Women Scientists: Reflections, Challenges, and Breaking Boundaries* (Oxford: Oxford University Press)
[20] Women *Princetonia*, accessed April 9, 2017 https://princetoniana.princeton.edu/history/women
[21] Howes R H and Herzenberg C L 1999 *Their Day in the Sun: Women of the Manhattan Project* (Philadelphia, PA: Temple University Press)
[22] McCaughey R A 2003 *Stand, Columbia: A History of Columbia University* (New York: Columbia University Press)
[23] Howes R H and Herzenberg C L 2015 *After the War: Women in Physics in the United States* (Bristol: IOP Publishing)
[24] Rosenberg R 2004 *Changing the Subject: How the Women of Columbia Shaped the Way We Think About Sex and Politics* (New York: Columbia University Press)
[25] Columbia University *Array of Contemporary American Physicists* accessed April 9, 2017 http://history.aip.org/history/acap/institutions/inst.jsp?columbia
[26] Wu C S 1973 Discovery of Parity Violation in Weak Interactions: Discovery Story I *Adventures in Experimental Physics Vol. γ*, ed; Bogdan Maglich (Princeton, NJ: World Science Education) 101–18
[27] Rainwater L J, Havens W W Jr, Wu C S and Dunning J R 1947 Slow neutron velocity spectrometer studies I. Cd, Ag, Sb, Ir, Mn *Phys. Rev.* **71** 65–79
[28] Wu C S and Albert R D 1949 The beta-ray spectra of Cu^{64} *Phys. Rev.* **75** 315–6
[29] Wu C S and Albert R D 1949 The beta-ray spectra of Cu^{64} and the ratio of N + / N - *Phys. Rev.* **75** 1107–8
[30] Wu C S and Shaknov I 1950 The angular correlation of scattered annihilation radiation *Phys. Rev.* **77** 136
[31] Boehm F and Wu C S 1954 Internal bremsstrahlung and ionization accompanying beta decay *Phys. Rev.* **93** 518–23
[32] Glauber R J, Martin P C, Lindqvist T and Wu C S 1956 Relativistic and screening effects in electron capture *Phys. Rev.* **101** 905
[33] Hughes V W, Marder S and Wu C S 1957 Hyperfine structure of positronium in its ground state *Phys. Rev.* **106** 934–47

[34] Biavati M H, Nassiff S J and Wu C S 1962 Internal bremsstrahlung spectrum accompanying 1S electron capture in decay of Fe^{55}, Cs^{131}, and Tl^{204} *Phys. Rev.* **125** 1364–72
[35] Lee Y K, Mo L W and Wu C S 1963 Experimental test of the conserved vector current theory on the beta spectra of B^{12} and N^{12} *Phys. Rev. Lett.* **10** 253–8
[36] Myneni K Symmetry destroyed: the failure of parity CCREWeb.org last modified on January 10, 2008 http://ccreweb.org/documents/parity/parity.html
[37] Siegbahn K (ed) 1955 *Beta- and Gamma-Ray Spectroscopy* (Amsterdam: North-Holland)
[38] Yang C N 1964 The law of parity conservation and other symmetry laws of physics *Nobel Lectures Physics 1942–1962* (Amsterdam: Elsevier)
[39] Lee T D and Yang C N 1956 Question of parity conservation in weak interactions *Phys. Rev.* **104** 254–8
[40] Wróblewski A K 2008 The downfall of parity—the revolution that happened fifty years ago *Acta Phys. Polonica* B **39** 251–64
[41] Yang C 1982 The discrete symmetries P, T, and C *J. Phys. Colloq.* **43** C8-439–C8-451
[42] Telegdi V L interview by Sara Lippincott, Pasadena, CA, March 4 and 9, 2002, *Oral History Project of the California Institute of Technology Archives* http://resolver.caltech.edu/CaltechOH:OH_Telegdi_V
[43] Interview of R Garwin by W Patrick McCray on June 7, 2001, Niels Bohr Library & Archives, American Institute of Physics, College Park, MD USA, www.aip.org/history-programs/niels-bohr-library/oral-histories/24292
[44] Wu C-S 1996 Parity violation *History of Original Ideas and Basic Discoveries in Particle Physics* **352** 381–400
[45] Hammond R 2010 *Chien-Shiung Wu: Pioneering Nuclear Physics* (New York: Chelsea House)
[46] Lederman L with Teresi D 1993 *The God Particle* (New York: Dell)
[47] Garwin R L, Lederman L M and Weinrich M 1957 Observations of the failure of conservation of parity and charge conjugation in meson decays: the magnetic moment of the free muon *Phys. Rev.* **106** 1415–7
[48] Friedman J I and Telegdi V L 1957 Nuclear emulsion evidence for parity nonconservation in the decay chain π^+—μ^+— e^+ *Phys. Rev.* **106** 1681–2
[49] Goudsmit S 1973 A reply from the editor of the physical review *Adventures in Experimental Physics Vol. γ*, ed Maglich Bogdan (Princeton, NJ: World Science Education) 137
[50] Franklin A 1986 *The Neglect of Experiment* (Cambridge: Cambridge University Press)
[51] Lee T D and Yang C N 2005 Commentary on parity nonconservation and a two-component theory of the neutrino *Selected Papers (1945-1980) with Commentary* ed Yang Chen Ning (Hackensack, NJ: World Scientific) 35–7
[52] Schmeck H M Jr 1957 Basic concept in physics is reported upset in tests *New York Times (1923-Current File)* January 16, 1957 **1**
[53] 2 Nobel winners have quiet party *New York Times (1923-Current File)* November 1, 1957, 8
[54] The Nobel Prize in Physics 1990 *Nobel Prizes and Laureates*, accessed April 9, 2017 https://www.nobelprize.org/nobel_prizes/physics/laureates/1990
[55] Hargittai M and Hargittai I 2004 *Candid Science IV: Conversations with Famous Physicist* (London: Imperial College)
[56] Winston R and Telegdi V L 1961 Fast atomic transitions within μ-mesonic hyperfine doublets, and observable effects of the spin dependence of muon absorption *Phys. Rev. Lett.* **7** 104–7

[57] Telegdi V 1973 Discovery of parity violation in weak interactions: discovery story III *Adventures in Experimental Physics Vol. γ* ed Maglich Bogdan (Princeton, NJ: World Science Education) 131–5
[58] Quinn H R, Deken J M, Loew G A and Prescott C Y memorial resolution: Edward Lee Garwin (1933-2008) *Stanford Report* June 17, 2009 http://news.stanford.edu/news/2009/june17/memres_garwin-061709.html
[59] Nobel A B will, translated as *Full Text of Alfred Nobel's Will* Nobel Media AB 2014. Web. 19 May 2017. http://www.nobelprize.org/alfred_nobel/will/will-full.html
[60] The Noble Prizes in the new century: An interview with Ralf Pettersson, Director of the Stockholm Branch of the Ludwig Institute for Cancer Research, the Karolinska Institute, and former chairman of the Nobel Prize Committee for Physiology/Medicine 2001 *EMBO Reports* **2** 83–5
[61] The Nobel Prize in Physics 1980 *Nobel Prizes and Laureates*, accessed April 9, 2017 https://www.nobelprize.org/nobel_prizes/physics/laureates/1980
[62] Noether E 1918 Invariante Variationsprobleme *Gott. Nachr.* **1918** 235–57
[63] Jiles D 2016 *Introduction to Magnetism and Magnetic Materials* 3rd edn (Boca Raton, FL: CRC Press)

Chapter 5

Maria Mayer

5.1 The seventh generation

Friedrich Göppert was a professor. So was his father, and his father's father, and his father's father's father, and so on back for five generations; Friedrich was the sixth. In Germany in the early 20th century, there was no reason there shouldn't be a seventh generation of Göppert professors.

Except that, while Friedrich and his wife Maria Wolff Göppert planned to have many children, all were stillborn but one [1]: Maria Gertrud Käte Göppert [2], born in 1906. If Friedrich was to pass his family profession on to a seventh generation, it would be his daughter who would need to pick up the mantle.

Maria's situation was far more favorable than that of Curie, Meitner, Payne, or even Wu. Her father knew the ropes, and had connections in academia. Like Wu, she was born at the perfect time—much earlier, and she would have seen the kinds of delays faced by Meitner; a bit later, and she would have had to face the misogyny of the Nazis.

Friedrich Göppert was not a feminist in the modern sense. He wished his daughter had been born a boy, and told her not to grow up to be a woman, a sentiment Maria transformed in to the advice often incorrectly attributed to Friedrich: 'don't be *just* a woman' [3].

Friedrich was also of the opinion that 'the mother is the natural enemy of the child', stifling the explorations of their progeny.

Maria, therefore, grew up as (in modern parlance) a 'free-range child', and as something of a tomboy [1].

By the time Maria was a teenager, women had been being admitted to German universities for a decade, but getting the necessary education was still not easy. Maria attended a private school meant to prepare girls for the *abitur* (equivalent to the Austrian *matura*), but the runaway inflation afflicting Germany at the time took its toll, and the school closed its doors one year before Maria had completed her course of study. Rather than finish her preparation elsewhere, she decided to take the *abitur* a year early, despite being underage.

Using her connections, she gained permission to take the exam at the age of 17. She and the girls a year ahead of her at her school all passed, but almost none of the boys she took it with did, a tribute to the high quality of education they had received as well as the talents of the girls [4].

Maria was admitted to the University of Göttingen, where her father taught. While it was her hometown university, it was also filled with the famous and soon-to-be famous: in mathematics, that included the venerable David Hilbert and the dramatically underpaid Emmy Noether, both of whom made key contributions to physics. In fact, Mayer, like Wu, began her university career majoring in mathematics. In physics, the future Nobel laureate Max Born had arrived just a few years before to take the chair of theoretical physics; James Franck took over the chair of experimental physics at the same time and won his Nobel while Maria was still a math major. Werner Heisenberg was a young member of the faculty. During Maria's time at Göttingen, Born's graduate students included J Robert Oppenheimer as well as Max Delbrück, who would go on to become an assistant to Lise Meitner before eventually switching to biology and winning a Nobel Prize in medicine. Fermi and Pauli had both recently spent time as postdocs at Göttingen, and Edward Teller arrived just after the end of Maria's time there [5].

Through her father, Maria often met these luminaries socially before she took classes from them or worked with them in the course of her studies [6]. While still in high school, for instance, Hilbert invited her to be the 'guest of honor' at a lecture of his [1].

Her biographer Dash provides a particularly interesting description of Maria in these early years. She was, by her own description, a 'brat'. She knew the ins and outs of university life, and was well-acquainted with all the movers and shakers at the University, and so could get away with a lot; after attending two classes of a course in psychology, for example, she convinced her professors to count her as if she had attended all semester.

The University of Göttingen was predominately male, but not exclusively so. In addition to Emmy Noether in mathematics, Hertha Sponer was a member of the physics faculty. While Lise Meitner was in Berlin, not Göttingen, James Franck was one of her closest friends, [7] so it is very likely that Maria met Meitner at some point during this span. But Maria seemed to prefer the society of men, and formed no special attachments to any of these female scientists [6].

While both Payne and Wu switched majors early in their college careers, Maria had been a mathematics student for three years before Born invited her to join his seminar, apparently on a hunch [1].

The hunch paid off—Maria discovered that she preferred theoretical physics to mathematics, and changed her course of study [4]. Soon she was Born's doctoral student—it wasn't always necessary to have a university diploma as a discrete step in those days, if the doctorate was the goal, much as in America today a master's degree is not always necessary *en route* to the PhD.

Oppenheimer was also in Born's seminar with Maria, and, although actually a little older than Maria, was seen as an *enfant terrible*—brilliant, but also lacking in tact or a sense of propriety. In seminar, he would often interrupt others—even Born—in order to provide a 'better' explanation of the topic at hand. Born, a mild-mannered professor, was unable to control Oppenheimer, so Maria took things into her own hands; she wrote a petition threatening to boycott the class unless Oppenheimer was reined in, and got most of the students in the seminar to sign it. Born seized the opportunity, leaving the petition on his desk in such a way that Oppenheimer was sure to find it while Born was out of the room. The warning had its desired effect; Oppenheimer stopped interrupting during seminar [8].

During that same year, Maria's father died unexpectedly, causing her to rededicate herself to completing the PhD [9].

It is clear that Maria's attitude toward physics was rather different than that of the other women featured in this book. For Meitner, physics was as necessary to her well-being as food; once she had tasted it, she could not imagine living without. For Payne, astrophysics was an itch, immediately drawing her in to long, sleepless nights contemplating its marvels or studying its secrets. Wu considered teaching early on, but it is clear that her heart was drawn to physics.

But Maria, while finding physics intellectually stimulating, and while she was extremely good at it (particularly the new science of quantum mechanics), could clearly have walked away without many regrets. Unlike the others, and unlike nearly all other women of the time, she was *expected*, from a fairly early age, to pursue a PhD. This was not an unpleasant prospect for her, but it was not a passionately held dream.

In 1928, Maria took a break from Göttingen to spend a term at Cambridge, a few years after Payne had graduated from there. Maria then returned to Göttingen to continue her work on her doctorate [10].

She had not been back in Göttingen long when Joe Mayer, an American post-doc, showed up at the house she shared with her mother, asking if they had a room for rent. Joe, eager to try out his German, spoke in that language, while Maria, fresh from Cambridge, replied in English. They soon fell in love (although Joe perhaps sooner than Maria), and within a few years were married.

Despite her 'sacred obligation' to complete her PhD in her father's memory, Maria did not approach its completion with zeal [6]. This was not for lack of encouragement. Her mother encouraged her, Joe pushed her, and one professor even locked her in his guest room until she had produced a rough draft [1]. The walls of the guest room were adorned with the signatures of scientists who had stayed there, including Einstein [6]. Maria added her name to the wall [5].

By early 1930, however, she managed to finish, and the couple moved to Baltimore, where Joe had been offered a position at Johns Hopkins.

5.2 From nuisance to necessary

Up until this point, Maria's gender had provided little barrier to her advancement. If she had not married Joe, and had remained in Germany, it is quite likely that she would have become a professor, and carried on the family tradition. But in the United States, many research universities still had not hired their first female professor. In addition, they possessed 'anti-nepotism policies'; ostensibly these were to prevent a supervisor from hiring a relative, but in practice universities often interpreted them as prohibiting a married couple from both being employed in any capacity [6]. Or, more specifically, as prohibiting a woman from being employed at the same institution as her husband; when Cecilia Payne married Sergei Gaposchkin, he seems to have had little trouble finding a paid position at the Harvard College Observatory.

Maria Mayer, as she was now known, helped earlier in her career by being the daughter of a university professor, was now stymied because her husband was one.

If she had possessed a greater ambition for such a role, she could have sought a faculty position at another institution. Wu and her husband, for example, spent much of their careers working in different cities, even while they raised a child. But Maria never seriously considered that option.

Still, her treatment at Johns Hopkins must have felt demeaning, even insulting. As a theorist, she did not need laboratory space, just an office. But instead of an office they offered her a workspace in the attic of the physics building! [1].

But just because she did not have a proper office and was not paid by the University did not mean that Johns Hopkins did not take advantage of her skills. She taught graduate courses, although she was listed in the catalog not by her full name, as other instructors were, but only with the mysterious moniker 'G', presumably short for her maiden name of Göppert [6]. Remarkably, she even served as dissertation advisor to a doctoral student [3].

In addition to teaching and mentoring, Maria maintained an active research agenda, travelling back to Göttingen in the summers and working with Johns Hopkins faculty during the year [3]. The result was a steady stream of papers, some in English, and some in German. She also began work, with Joe, on a textbook on statistical mechanics.

Joe had been promoted from 'associate' to 'associate professor' in 1937 [11], but was not tenured. Johns Hopkins was facing financial difficulties, and when a new president came in he brought with him an era of cost-cutting [12]. In 1939, Joe lost his job.

Was Joe 'cut', or was he 'fired'? In other words, was he a victim of downsizing of the chemistry department, vulnerable because of the lack of the tenure, or was he targeted for removal? The Mayers had no doubt that it was the latter, but were never quite sure why. They suspected that it might have something to do with Maria: perhaps misogyny, or anti-German sentiment [1]. Strange as it may seem, the work that Maria did for free for the university was seen as a burden by some, requiring concessions such as giving up part of the attic to be her office.

Joe's next stop was Columbia, where he was hired as an associate professor, this time with tenure [13]. Once again, anti-nepotism rules prevented Maria from being paid, or from having a regular position. With the completion of their statistical mechanics book in 1940 [14], however, it was imperative that Maria have a title to put by her name for the title page. While the Columbia physics department did not see fit to give her even an honorary position, one of the professors, Harold Urey, managed to get permission for her to give a few lectures in the chemistry department, allowing her to be credited on the title page as 'lecturer' [1].

The year after the textbook was published, one notable paper by Maria provided a theoretical justification for placing the actinides beneath the lanthanides on the periodic table, rather than as another row of transition metals [15]. While experimental evidence had been accumulating in favor of this arrangement, Mayer's paper employed her knowledge of both mathematical quantum mechanics and chemistry—an unusual combination—to place this on a firm footing, giving the periodic table the shape we're familiar with today.

On 7 December, 1941, the Japanese attacked Pearl Harbor, bringing the United States into World War II. This set in to motion a sequence of events which would change the trajectory of Maria's career.

First, Sarah Lawrence College, a liberal arts college for women just north of New York City, lost one of its mathematics professors to war work, creating an immediate need for an instructor for the spring semester [16]. Maria was a *logical* choice, as she was available on short notice, but it was clear neither to Maria nor to the search committee at Sarah Lawrence that she would be a *good* choice. She had never taught undergraduates before, and was certainly not familiar with the progressive style of education, inspired by the works of John Dewey, that was employed at Sarah Lawrence. The search committee also couldn't quite get their heads around her field of expertise, with one (correctly) identifying her as a theoretical physicist, while another described her as a physical chemist, a label more appropriate for her husband [17]. She and the college quickly warmed to each other, however (figure 5.1).

Figure 5.1. Maria Mayer (second from right) in the Faculty Dining Room at Sarah Lawrence College. Copyright Sarah Lawrence College Archives.

Sarah Lawrence was clearly pleased with the results of that first semester, because in June they increased her teaching responsibilities to two courses each term for the next academic year, while only requiring her to be on campus three days a week so that she could continue to teach occasional lectures at Columbia and also spend time with her children. She was also given students to advise. Since Sarah Lawrence expected students to remain with the same academic advisor throughout their four undergraduate years, this suggested that the college expected her to remain for the long term [17].

At Sarah Lawrence, she faced none of the barriers she had encountered at Johns Hopkins and Columbia. Since Joe did not work there, anti-nepotism rules did not apply. And since it was a women's college, there were already women on the faculty. At the time, Sarah Lawrence did not have tenure [17], and to this day it does not have rank (assistant, associate, etc). Sarah Lawrence's egalitarian system meant that, from Maria's first day on campus, her formal status earned a significant paycheck for her efforts. At Johns Hopkins and Columbia she had occasionally received small sums, perhaps paid directly by a sympathetic member of the faculty; Joe recalls at Johns Hopkins this amounting to perhaps $100 per year (under $2000 in 2017 dollars) [18]. In her second academic year at Sarah Lawrence, she was paid $2100 [19], or $31 000 in 2017 dollars. While not a princely sum, it was at least a real salary, particularly considering the work was not quite full-time [17].

From 1942, Joe was called to Aberdeen Proving Ground in Maryland to consult for the Army, working there four days a week. Eventually, his work was to take him to the Pacific theater of the war, leading to a long time away from home.

In the fall of 1943, it was Maria's turn to be called to serve her country. The Manhattan Project was well underway, and Harold Urey, the Columbia professor who had argued for a position on her behalf, was in charge of finding a way to separate the isotopes of uranium. In September, Urey wrote President Constance Warren of Sarah Lawrence, explaining that Maria was needed for the war effort (figure 5.2). She was soon granted leave from the college, and began working at Columbia. While Maria secured Urey's agreement that the position would be part-time, in practice a scientific investigation of that importance and urgency never is, particularly as the research team she supervised grew to include more than fifty scientists [8]. With Joe at Aberdeen four days a week and money no longer as much of an issue, the Mayers hired a nanny to care for the children [17].

Wartime secrecy was such that Joe couldn't tell Maria what he was working on (among other things, proximity fuses for torpedos), and she couldn't tell him (atomic bombs). Joe had some sense of it, however, telling Maria that 'I'm working on this war, you're working on the next' [1].

Urey needed all the scientists he could get, and Maria recommended one of her former students from Sarah Lawrence, Susan Chandler Herrick, for a position, a proposition which Urey quickly agreed to. Herrick had been the only student in Maria's physical chemistry class the year before, and now had a year off between Sarah Lawrence and medical school [18].

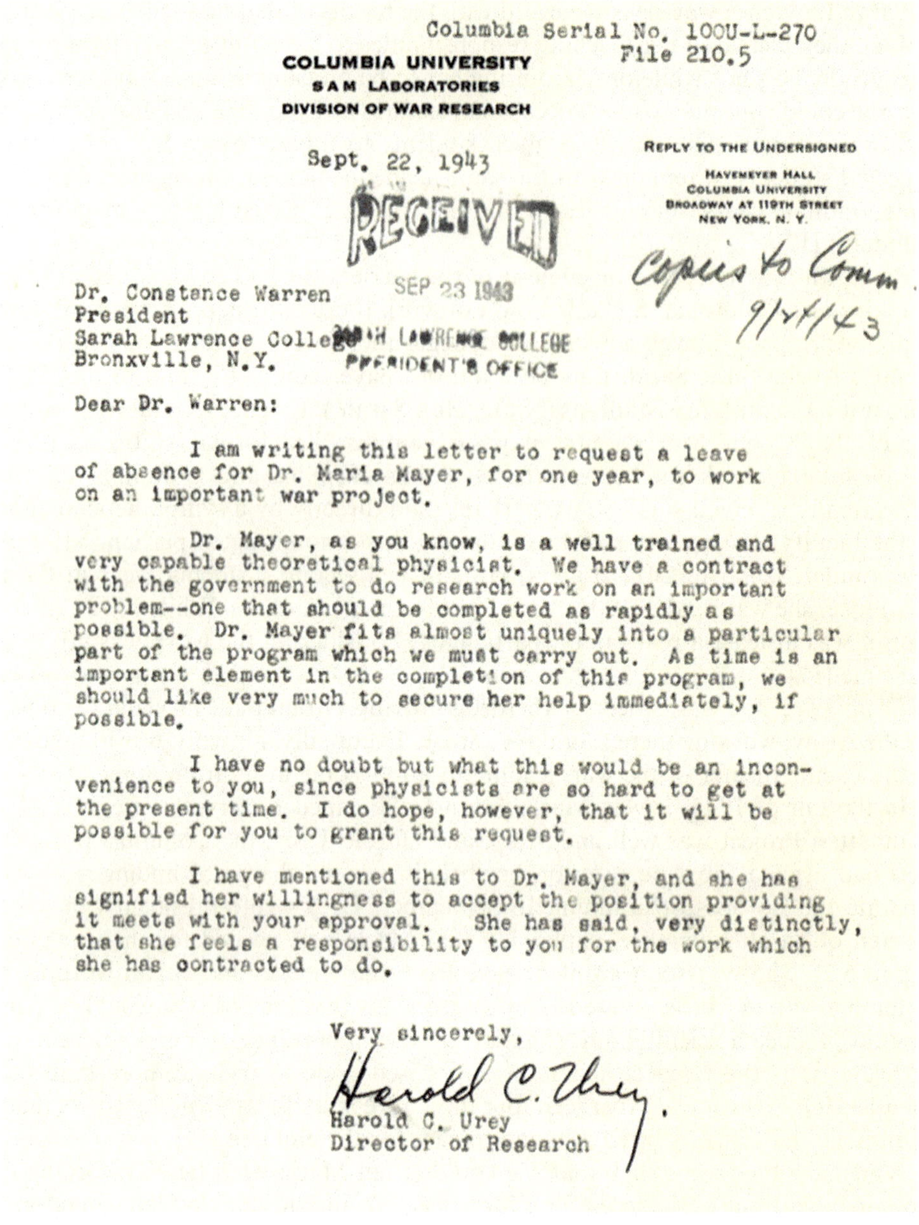

Columbia Serial No. 100U-L-270
File 210.5

COLUMBIA UNIVERSITY
SAM LABORATORIES
DIVISION OF WAR RESEARCH

REPLY TO THE UNDERSIGNED
HAVEMEYER HALL
COLUMBIA UNIVERSITY
BROADWAY AT 119TH STREET
NEW YORK, N. Y.

Sept. 22, 1943

RECEIVED
SEP 23 1943
SARAH LAWRENCE COLLEGE
PRESIDENT'S OFFICE

Copies to Comm.
9/24/43

Dr. Constance Warren
President
Sarah Lawrence College
Bronxville, N.Y.

Dear Dr. Warren:

I am writing this letter to request a leave of absence for Dr. Maria Mayer, for one year, to work on an important war project.

Dr. Mayer, as you know, is a well trained and very capable theoretical physicist. We have a contract with the government to do research work on an important problem--one that should be completed as rapidly as possible. Dr. Mayer fits almost uniquely into a particular part of the program which we must carry out. As time is an important element in the completion of this program, we should like very much to secure her help immediately, if possible.

I have no doubt but what this would be an inconvenience to you, since physicists are so hard to get at the present time. I do hope, however, that it will be possible for you to grant this request.

I have mentioned this to Dr. Mayer, and she has signified her willingness to accept the position providing it meets with your approval. She has said, very distinctly, that she feels a responsibility to you for the work which she has contracted to do.

Very sincerely,

Harold C. Urey

Harold C. Urey
Director of Research

Figure 5.2. Harold Urey's letter to President Warren. Copyright Sarah Lawrence College Archives.

Soon after Maria began her work with Manhattan Project, she wrote President Warren to assure that she would be able to return to Sarah Lawrence when she was done:

> It is very sad for me to interrupt my work at Sarah Lawrence, but I hope it is only temporary. The contact with the girls at the college has been very pleasant

indeed and each year I have enjoyed my work more. Teaching and the things it brings with it, namely, the human contact with a group of eager and interested young girls, is a wonderful supplement to doing research work; all of which leads up to the fact that I hope to be back with you again.

The work I am going into is sponsored directly by the Army. They are in very great need of scientists and I believe that it is probably my duty to help out. I told Miss Doerschuk, however, that if my leaving now imperiled the possibility of my return to Sarah Lawrence, the personal sacrifice demanded is too much and I would rather try to fight my draft... [21].

By February, due to a combination of missing the teaching and mentorship she had done at Sarah Lawrence, ambivalence toward working on a bomb, and physical illness and stress, Maria wanted to return to Sarah Lawrence for the next academic year [17]:

My own desires have been clear to me for quite some time. I would like to return to my work at Sarah Lawrence. For one thing, spending at least 40 h a week in the laboratory is rather strenuous when combined with two children. But more than that, I have missed very much the teaching as it is done at Sarah Lawrence—not the mere imparting of knowledge, but the human contact with developing personalities [22].

Soon, a tug-of-war developed between Sarah Lawrence and the Manhattan Project. In April, Sarah Lawrence offered her a *full-time* position at Sarah Lawrence, at a salary of $2800 ($39 000 in 2017 dollars) [23]. Urey countered by asking Maria to work for him for 1.5 days a week, a request Maria refused. She eventually agreed to spend half a day a week at Columbia supervising a graduate student working on the project, while working full time at Sarah Lawrence [17]. In December Sarah Lawrence raised her salary again, this time to $3000 ($41 000 in 2017 dollars) [24]. In just a few years, Maria had gone from being unable to find a paid job anywhere, to being a full-time member of the faculty at Sarah Lawrence and a paid consultant at Columbia!

As a full-time member of the faculty, Maria participated eagerly and actively in the full life of the college. She was often asked to serve on panels related to the war; on one such panel in February of 1945 she ominously warned 'man's scientific discoveries and inventions might very likely destroy him', a quote printed in the next issue of the school paper [25].

She also developed a new 'humdinger of a course', [26] entitled Fundamental Physical Science, which would provide a unified approach to science from a liberal arts perspective:

The course presents man's knowledge of the Universe and the atoms, which compose it. It deals, consequently, with subjects, which are basic to the sciences of astronomy, geology, chemistry and physics. Science is treated as a liberal art rather than pre-professional training. The course is, however, prerequisite for further study in either physics or chemistry. The laboratory work contains chemistry and physics as well as observation of stars. No previous preparation in mathematics or science is required [27].

Initially, President Warren was sceptical, asking 'is it more important that these kids should know these high-faluting things than to be able to regulate a flu on a furnace?'

Maria, with her long experience with academic administrators, knew just how to respond to Warren, who had been trained in the humanities, 'do we teach them English so that they can read a cookbook?' [26].

The course proposal was approved.

5.3 A new era

Following the success of her new course, Maria was given yet another raise for the 45–46 school year, this time to $3200 per year ($43 000 in 2017 dollars) [28]. While still not an exorbitant salary, it was becoming a good one, and its rapid and repeated increase suggests how much the college valued her.

The Manhattan Project valued her as well, and while Columbia remained her home base for that work, she sometimes made trips out to Los Alamos, and was eventually given the use of a house there so she would have a comfortable place to stay during her visits [13].

Maria was at Los Alamos when Joe returned to the US from his trip to the Pacific theater; they managed to rendezvous in New Mexico and then return together to their home in New Jersey. Maria had hoped to be at Los Alamos when the first bomb was tested, but she did not know when that would be; as it turned out, the test happened not long after they got back [1].

Every aspect of the project was, of course, revealed only on a need-to-know basis. Neither Urey nor Maria knew the day of the first test, although they both deduced it when on 16 July Urey was unable to reach anyone at Los Alamos by phone.

Joe didn't know what his wife was working on, and did not press her for details, but he could surely guess the general contours. So, she later recalled, did her inquisitive students at Sarah Lawrence:

> I taught them a lot of nuclear physics but I never mentioned uranium…I couldn't. But a number of them discovered it and said, 'Why don't you talk about it?' I said, 'I know nothing about it.' They said, 'Oh, you aren't allowed to, you are working on it.' I said, 'No, I just don't know anything about it.'… And…after the bomb fell, they said they knew, they had known, and they gave lectures to the rest of the class to explain how the atomic bomb worked [26].

Maria and Joe were on vacation with their children in Nantucket on 6 August, the day the atomic bomb was dropped on Hiroshima. The bomb used on Hiroshima used uranium, and was thus most directly connected with Maria's work on isotope separation. (The bomb used on Nagasaki three days later employed plutonium created in a nuclear reactor developed under Fermi's leadership.)

The Mayers were walking on the beach in Nantucket when they got the news [12]. With the secret out, Maria filled Joe in on what she had been doing at Columbia and Los Alamos.

Maria's star was ascending. She had a reliable faculty job, connections with the leading physicists and chemists in America, and the cachet of having worked on the Manhattan Project.

The Columbia chemistry department, however, despite Urey's enthusiastic support of Maria, was unmoved. For some time, Maria had been in the habit of attending the weekly seminars held by the chemistry department. Before the war, she and Joe would go with Professor George Kimball, another faculty member, and Alice Kimball, his wife. Alice had a degree in chemistry from MIT, and, much like Maria, orbited the chemistry department without really being a part of it, assisting her husband with his studies on campus. Like Joe, George Kimball was called away for war work, and, like Maria, Alice was recruited by Urey to help at Columbia with the Manhattan Project [29].

Maria and Alice kept going to the department seminars on their own after their husbands got called away. This was tolerated. But following each seminar, the faculty would go out to dinner together which was, of course, where much of the most interesting science would be discussed. Maria was told by a member of the chemistry department that the presence of herself and Alice at the dinners was 'awkward', and that they should limit themselves to the seminars proper. Faced with that restriction, Maria stopped going to either [1].

(C S Wu had arrived at Columbia in 1944, just as Maria was returning to full-time work at Sarah Lawrence. It is likely they met, but with Maria's work at Columbia limited to a half day a week working with Teller's graduate student, and with her withdrawal from the life of the department, they did not cross paths often.)

The shortsightedness of first Johns Hopkins, and then Columbia, cost both institutions. Not just in that they lost Maria, but also a good many other physicists and chemists of great prominence. James Franck, the chair of experimental physics at Göttingen when Maria was a student and a good friend to Lise Meitner, left for Johns Hopkins when Hitler came to power. But Johns Hopkins, in turn, lost him to the University of Chicago in 1938.

Fermi had come to Columbia in 1938, but his work on the Manhattan Project took place at the University of Chicago, where he built the world's first nuclear reactor. In the summer of 1945, Chicago lured him away from Columbia for good. Urey and Teller were also recruited. And so was Joe [1].

Writing to Harold Taylor, the new president of Sarah Lawrence, Maria said that she was 'very sorry to leave, but I have no choice in the matter' [30]. As when she was working on the Manhattan Project, Sarah Lawrence was loath to see her go, and successfully negotiated with her to stay for one more semester while they found a replacement. Taylor wryly observed that, in terms of searching for a replacement, 'now that you scientists have succeeded in administering the final blow to the old world perhaps we can have some of the young physicists back in the academic halls' [31].

Chicago, for its part wanted Maria as well. After Joe accepted the position of full professor, they offered Maria a position as associate professor. As Maria said, it was the first institution where she 'was not considered a nuisance, but greeted with open arms' [9][1]. Open arms, yes, but not open wallets; due to nepotism rules, she was to be a 'voluntary' associate professor, meaning that she would do the work without pay.

5.4 Magic

At Chicago, both Maria and Joe were made part of the Institute for Nuclear Studies; unlike her position at the university, this position did provide a salary for Maria. Up until that point, Maria had not thought of herself as a nuclear physicist: her specialties were chemical physics, quantum mechanics, and applied mathematics. Although she had worked on uranium separation during the war, that work had little to do with what was going on inside the nucleus itself. But she gamely began to learn the field, relying particularly heavily on Teller to teach her the ropes. Teller, in turn, valued Maria's skill with advanced mathematics. And both found their collaborations to be very productive (figure 5.3) [1].

Teller became interested in the origin of the chemical elements, and got Maria interested in it as well. As they worked on the theory, she began to notice that certain quantities of neutrons and protons formed especially stable nuclei. They were not the first to notice this, or to try to explain it; there had been particular interest in the phenomenon in the 30s, starting soon after the neutron was discovered [32].

That initial interest had led to a flurry of theorizing, attempting to explain these 'magic numbers' for neutrons and protons by shell models analogous to those used to explain the energy levels of electrons and the structure of the periodic table (see the Science Summary at the end of this chapter). But the success of Meitner and Frisch in explaining fission by using Gamow and Bohr's liquid drop model largely ended that line of inquiry. If the neutrons and protons in a nucleus moved randomly in the manner of molecules in a liquid, the thinking went, then it was surely nonsensical to think of them as having well-defined orbitals organized into shells. The magic numbers themselves became regarded as uninteresting, a numerical curiosity [33][2]. In fact, Maria, new to nuclear physics, did not even know about the earlier investigations until they were mentioned to her by a colleague. Once she learned of them, she made a careful search of the literature to see what other work, both experimental and theoretical, had been done in the area [1].

The early evidence for magic numbers was not overwhelming, allowing for the possibility that they were due to physicists seeing patterns where none existed.

---

[1] In saying this, Maria was not being entirely fair to Sarah Lawrence. Sarah Lawrence did not *greet* her with open arms, but from the start considered her a necessity rather than a nuisance. By the time she had taught there for a few years, she was a valued, even loved, member of the college faculty, and the students, faculty, and staff were truly sorry to see her go.

[2] That may seem peculiar, but it is not without precedent. Consider the spacing of the planets in the solar system. In the 18th century, Johann Titius and Johann Bode discovered a mathematical 'law' that, to a good approximation, matched the measured spacing. This law even led to correct predictions for where newly-discovered planets would be found. But then the law failed for the most distant planets, and was demoted to the status of a mathematical curiosity. Recent work, however, has suggested that there may be a theoretical basis to it after all, related to the formation of orbital resonances, and that laws of the Titius–Bode type may be a common feature of planetary systems. More than two centuries after its initial formulation, the jury is still out.

Figure 5.3. Left to right: Edwin Teller, Maria Mayer, Joe Mayer, and James Franck. Otto Stern photograph collection, BANC PIC 1988.070:066–PIC. Courtesy of The Bancroft Library, University of California, Berkeley.

In fact, Teller tried to talk her out of pursuing the matter further:

> That seemed like a detail to me, but Maria thought that the regular repetitious appearance of these abundances must have an interesting explanation in itself; whether it was connected with the origin of elements was not the issue. I persisted in disparaging her interest until finally she lost her temper [34].

But Mayer added more recent data to the mix, and found that the case had been considerably strengthened [3]. In 1948 she published her results, including the bold statement that 'the complete evidence for [magic numbers] has never been summarized, nor is it generally recognized how convincing this evidence is' [35].

In fact, Teller had lost interest, as had almost everyone else. There were only two scientists who encouraged Maria in her investigation: Fermi, who had a hunch that Maria was on to something, and her husband Joe, who provided spousal support and a chemist's perspective [3].

The situation has parallels with Payne's dissertation work two decades earlier. Both Payne and Mayer were interpreting experimental data relevant to the relative abundance of elements (and, in Mayer's case, isotopes) in the Universe, and both had arrived at an 'impossible' result that contradicted known theory. In Payne's case, she found that hydrogen and helium were the predominant elements, throwing a wrench into Eddington's models and everyone who depended on them. Meitner found that the magic numbers appeared to be real anomalies, raising questions about the liquid drop model of the nucleus.

But, crucially, their life situations were very different, as were their personalities. In 1925 Payne was a *wunderkind*, eager to prove her worth, surrounded by scientists far more established than she. Mayer, on the other hand, was a competent but unremarkable mid-career scientist, secure in her reputation and position. At worst, she would risk being thought of as quixotic, but that was a description that could be applied to many scientists at a similar stage in their careers.

The result was that, unlike Payne, Mayer did not back off from her own conclusions. Instead, she set about trying to find a theoretical explanation.

5.5 A different way to win a race

But while Maria, Joe, and Fermi were the only scientists at Chicago to take an interest in the magic numbers, there was another group, unknown to them, who were also still thinking about the phenomenon. In Germany, Otto Haxel and Hans Suess were interested, and soon convinced Hans Jensen to think about the problem. Jensen kept the idea on the back burner until he saw Mayer's paper and had a discussion with Niels Bohr that indicated that Bohr took it seriously. Since Bohr was one of the key architects of the liquid drop model, Jensen felt that he should take it seriously as well, and began to search for an explanation in earnest [36]. Others, including Eugene Feenberg and Kenyon Hammack at Washington University and Lothar Nordheim at Duke, began attacking the puzzle with renewed intensity. Maria had set off a race, although she didn't realize it.

But Maria was changing. Always before, research in physics had been a sort of hobby for her, intellectually engaging and productive, but something she could regularly set aside for dinner with friends or a good game of bridge. Gradually, her work on the magic numbers became more and more intense. Bridge and dinners did not cease entirely, but became breaks from her primary passion, physics [1].

Maria continued to talk the problem over with Fermi in her office, and with Joe at home. During one of those conversations with Fermi, someone stopped by to let him know that there was a phone call for him in his office. As he rose to leave, Fermi threw out a suggestion: 'is there any indication of spin–orbit coupling[3]?' Maria's reaction was instantaneous: 'Yes, of course and that will explain everything'. Fermi left for his call, and ten minutes later Maria had worked out the solution. She had explained the magic numbers [32].

---

[3] See the Science Summary at the end of this chapter

That night, she told Joe that the magic numbers had been explained. Despite having listened to her progress every step of the way, or perhaps because of that, Joe did not immediately realize that she meant that she had explained them, rather than someone else. Once he did, he encouraged her to publish immediately.

Maria had, however, seen preprints of a paper from Feenberg and Hammack [37], and another by Nordheim [38]. While their proposed solutions to the problem were different than hers, she wanted to use some of their results to strengthen her analysis, and did not want to take advantage of having seen their preprints. After discussing the matter with the editors of *Physical Review*, it was decided that their full papers would appear in the same issue, along with shorter papers by Maria [39] and one by Feenberg, Hammack, and Nordheim [40] explaining the differences between the models. For some reason, *Physical Review* took its time publishing these papers; while the last of them was received by the end of February 1949, the issue was not published until 15 June.

By that time Jensen had independently found the same solution as Maria [41]. They had submitted a series of papers to the German journal *Naturwissenschaften* and to *Physical Review*. Their submission to *Physical Review* arrived in April [42]. *Physical Review* inexplicably failed to recognize that it covered the same topic as the papers scheduled for the 15 June issue, and published it on 1 June.

This sequencing likely hinged on the fact that *Physical Review* tended to turn around short papers (then called Letters to the Editor) more quickly than full-length articles. Maria's paper, if submitted as a standalone Letter to the Editor, would have been published before Jensen's, except that she had suggested it appear in the same issue as the full articles by Feenberg and Hammack and Nordheim.

Fortunately, the date received is printed at the beginning of published papers, thus establishing that Maria's paper had come in before Jensen's, even though it was published later. Everyone agreed that the discoveries were independent and complete. (This should be contrasted with Telegdi's weaker claim *vis a vis* Wu, since Telegdi's paper was not complete when he submitted it, and appeared to be rushed because he knew of the other results.)

Brief letters of the kind Maria and the Germans had written are traditionally followed by longer articles filling in the details. The Germans had already submitted slightly longer articles to *Naturwissenschaften* [43, 44].

Maria, though, hesitated. She wanted better experimental evidence; she wanted a mathematically sound theory.

Meanwhile, Fermi and, most persistently, Joe were urging her to write the follow-up articles. At one point their teenager daughter witnessed Joe physically putting a pencil in to Maria's hand to encourage her to write [13]. At times, Joe lost patience: 'For God's sake, write it up! You have to write it up—you have to do it right now!' [1].

She did, eventually, write it up, producing a pair of papers, one providing a careful compilation of the experimental evidence for the theory [45], and the other a quantitative theoretical basis for it [46].

Who, here, has priority? It is clear that Jensen and Maria developed the key ideas of their theory independently. Maria had a paper on the topic received first, but Jensen had one published first, and Jensen's more complete papers led Maria's by months

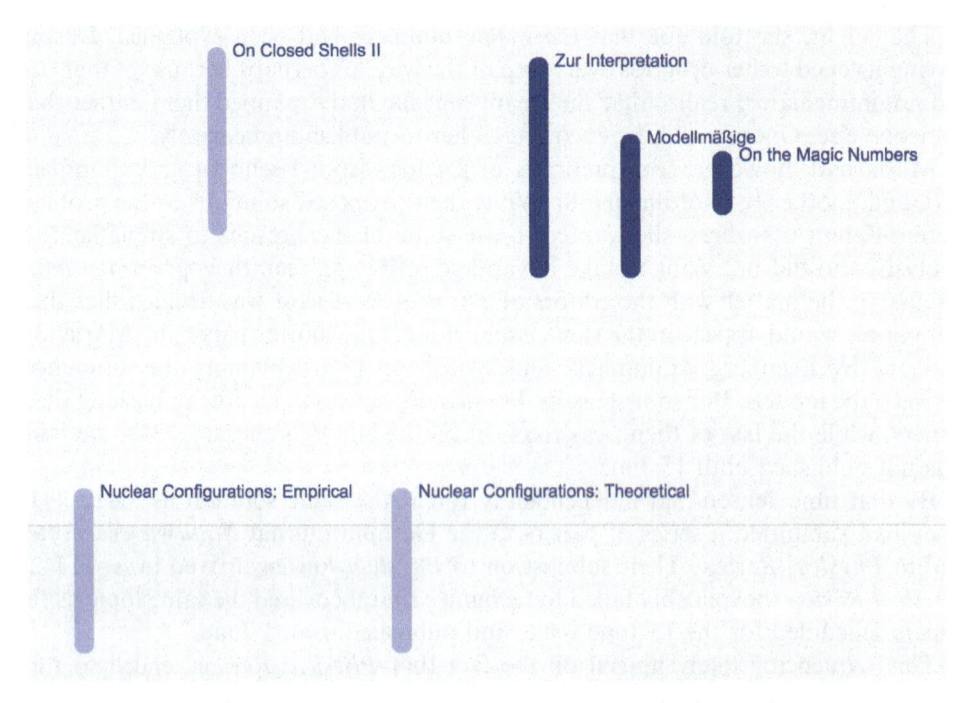

Figure 5.4. Schematic timeline of Mayer's and Jensen's 1949–50 papers on nuclear shells. Mayer's are the lighter shades on the left; Jensen's the darker on the right. Time progresses top to bottom, with each bar starting when the paper was received at the journal, and ending when it is published. Publication dates for the two German papers by Jensen are approximate. Mayer submitted her first paper earlier than any of Jensen's, but it was published after Jensen's first paper, and her remaining two, fleshing out the details, were much later than the remaining two from Jensen.

(figure 5.4). Maria's two papers in 1950, however, were more comprehensive than those of Jensen. One or the other could have easily become cantankerous, as Telegdi did when beaten by Wu, or bitter at unjust treatment, as Meitner did with Hahn.

Certainly, Maria's accomplishment was the more impressive of the two. Jensen had used the experimental evidence collected by Suess to formulate his theory [41]. Maria, while she had used Fermi and Joe as sounding boards, did all the hard work, both of analyzing experimental evidence and of constructing a mathematically sound theory, on her own.

Jensen and Maria, however, instead of becoming rivals, became acquainted with each other, and soon became friends and colleagues [3]. They decided to write a textbook together on their joint theory and the evidence for it. Writing this textbook was not easy—Jensen and Maria lived thousands of miles apart, and both found writing to be difficult at times [1]. But in 1955, their textbook was published [47]. Significantly, Mayer's name appeared first even though it would come after Jensen's alphabetically, suggesting that Mayer was the more important contributor.

The book was thorough and well-received. At this time, Maria and Jensen began to garner Nobel nominations. While Jensen's first publications on the topic included the names of Suess and Haxel, it was now clear that the theory belonged primarily to him and Maria, making a joint prize a possibility. From 1955 to 1963, the pair were nominated jointly 26 times [48, 49]. (Jensen received three solo nominations, all from the same person in different years, while Mayer received one.)

In 1959, the University of California was opening a new campus in La Jolla (eventually it would be renamed UC San Diego, for the larger nearby city). Joe and Maria were each offered the position of full professor. The University of Chicago saw the error of its ways too late, countering with an offer to finally pay Maria. That fall, Joe and Maria, joined their old friend and ally Harold Urey at the new institution.

That's not to say that she didn't have brushes with sexist treatment at UCSD as well. Initially, Maria was provided a full-time position at half pay, a policy which made no sense from the perspective of either fairness *or* anti-nepotism efforts. Within a few months, however, this was rectified, providing her an academic-year salary of $14 208 ($117 000 in 2017 dollars) [6].

In 1963, sixty years after Marie Curie had won a Nobel Prize in physics, Maria Mayer became the second woman to receive a Nobel in that field [50].

5.6 Quarter loafs

While it would have been nice to end this chapter on a positive note, there is one more issue I'd like to examine first [51].

Maria Mayer's Nobel prize was, as should be expected, shared with Hans Jensen. But their prize together was shared with Eugene Wigner, for work which was only loosely related. That meant that half the prize money went to Wigner, and then half of the remaining half went to Mayer; thus, she got a quarter-share. For a prize that can be shared by at most three people, this is an interesting, but not unprecedented, allocation.

The first time someone got a quarter-share of a Nobel Prize was Marie Curie, in 1903. She shared her prize with her husband, which they shared together with Henri Becquerel. The implication, in that case, was clear—the husband and wife team of the Curies was treated as if they were one person for the purposes of allocation, much as Maria and Joe Mayer were treated as one person by the University of Chicago in terms of remuneration. The Nobel Prize formally gave equal amounts to each of the Curies, while the University of Chicago formally gave the full amount to Joe and none to Maria, but the practical result was essentially the same.

Thus, women have won a total of two quarter-shares of a Nobel Prize in physics. During the period 1903–63, quarters-shares were not used even once as part of a trio of men in physics. The very next year, however, it was, with Charles Townes winning half the prize and Nicolay Basov and Aleksandr Prokhorov sharing the remainder.

In table 5.1, I've tabulated the fraction of quarter-shares by gender in each of the science categories of Nobel, both for the period 1901–63, and 1964–2016. Quarter-shares have never been awarded in literature, peace, or economics.

While the dramatic disparity between the number of Nobel Prizes awarded to men and women has not changed much in the years since Maria Mayer won

Table 5.1. Fraction of Nobel Prize winners in the science categories to receive ¼ shares, by gender, era, and category.

	1901–63 F	1901–63 M	1964–2016 F	1964–2016 M
Physics	2/2 = 100%	2/78 = 3%	No prizes	24/123 = 20%
Chemistry	0/2 = 0%	2/67 = 3%	0/2 = 0%	12/104 = 12%
Medicine	1/1 = 100%	3/84 = 4%	2/11 = 18%	10/108 = 9%
All Science Prizes	60%	3%	15%	14%

her prize, the disparity in the allocation of quarter-shares has essentially vanished[4], with the practice being used more often for men and less often for women than in the previous era.

5.7 A typical genius

Nobel Prizes are awarded (almost) every year. Unlike many scientific awards, they are not supposed to be given for a body of work across a career, but for 'the person who shall have made the most important discovery or invention within the field of physics' in the previous year [52]. The restriction to the previous year has long been disregarded as impractical, because it's often not clear what discovery or invention was most important until many years have passed, so that it is now interpreted as 'the year when the full impact of the discovery has become evident' [53]. Many Nobel Laureates, therefore, are excellent scientists who achieved a historic discovery or invention, but are not among the most prominent scientists of the past hundred and sixteen years when judged across the span of their careers. Thus Curie and Einstein and Fermi would still be considered extraordinary physicists even without the particular achievements that earned them their prizes, while Charles Barkla (discovery of x-ray spectra of elements), Charles Guillaume (invention of invar nickel-steel), and Jean Perrin (verification of the existence of molecules) were highly accomplished scientists who did particular work that merited the Nobel.

Mayer belongs more with the latter category than the former. She was brilliant and very gifted, particularly when it came to the new science of quantum mechanics. Her work with magic numbers was as good as it gets, and she fully deserved both her faculty position at UCSD and the Nobel Prize. But she did not have a career that matched Meitner's or Pauli's or any of the few dozen other physicists that led the field in the first six decades of the 20th century. She was, in that sense, typical of most Nobel laureates in the sciences.

In the context of the topic of this book, that is important because she was given the support these 'typical geniuses' need in order to make a discovery of historic importance. Einstein would have made his discoveries no matter what—in fact, his

---

[4] A chi-squared test on the 1903–63 data in table 5.1 reveals a significant difference between the rate of ¼-shares for men and women, while the difference in the 1964–2016 data is not statistically significant.

most famous work was done while he was working as a patent clerk, with little encouragement from the scientific community. Although it's harder for an experimentalist, Meitner, too, would have pursued physics no matter what, and made her most famous discovery outside of the laboratory, traipsing through the snow with her nephew as a refugee in Sweden. But most Nobel laureates in the sciences—meaning most male Nobel laureates, because most Nobel laureates are male—are not like that. A chance deviation early in their education, a sustained set of personal setbacks, or a lucrative job offer distracting them from a project could easily have sent them in another direction and prevented them from winning their prize.

Much has been written about why there are so many fewer female Nobel laureates in the sciences than male ones. It is likely due to a combination of causes, including implicit or explicit bias within the process that awards the prizes. But Mayer's prize shows what can happen when a woman of superior ability and average motivation receives crucial support along the way, from her parents, her educators, her co-workers such as Urey, Teller, and Fermi, and her husband. She then demonstrated what she was capable of, which was physics at the highest level. Historically, men are much more likely to be provided that kind of support than women, and so are much more likely to make discoveries of the highest caliber.

5.8 Science summary: nuclear shell model

First noticed soon after the discovery of the neutron, Mayer and Suess used a variety of techniques to analyze existing experimental data to find the number of neutrons or protons which made nuclei particularly stable: 'magic numbers'. In her 1948 paper, Mayer found evidence for the stability of nuclei with 50 or 82 protons and of nuclei with 50, 82, or 126 neutrons [35]. By the time of her Nobel lecture, the list of magic numbers for both neutrons and protons had expanded to include 2, 8, 20, 28, 50, 82, and 126 [32].

In the latter part of the 19th century, chemists had established the periodic table of the elements. The structure of the table is now known to be determined by the electrons orbiting the nucleus. Physicists such as Mayer and Jensen who developed shell models of the nucleus were thinking of them as analogous to the progress that had previously been made to understanding the structure of the periodic table, so I'll discuss it first, before moving inward to the nucleus.

In the 19th century, chemists such as Dmitri Mendeleev noticed that when elements were arranged in order of increasing atomic weight, similar properties reappeared in a similar order. For example, the sequence lithium, beryllium, boron, carbon, among light elements has similar properties to the sequence sodium, magnesium, aluminum, silicon, among heavier elements, with sodium being like lithium, magnesium like beryllium, and so on. Eventually it was realized that the sequence should be written in terms of the number of electrons in the neutral atom ('atomic number'), leading to our modern periodic table (figure 5.5).

The shape of the table is dictated by the rules of quantum mechanics. In any atom, each electron is required to move in a different manner. It is as if an instructor

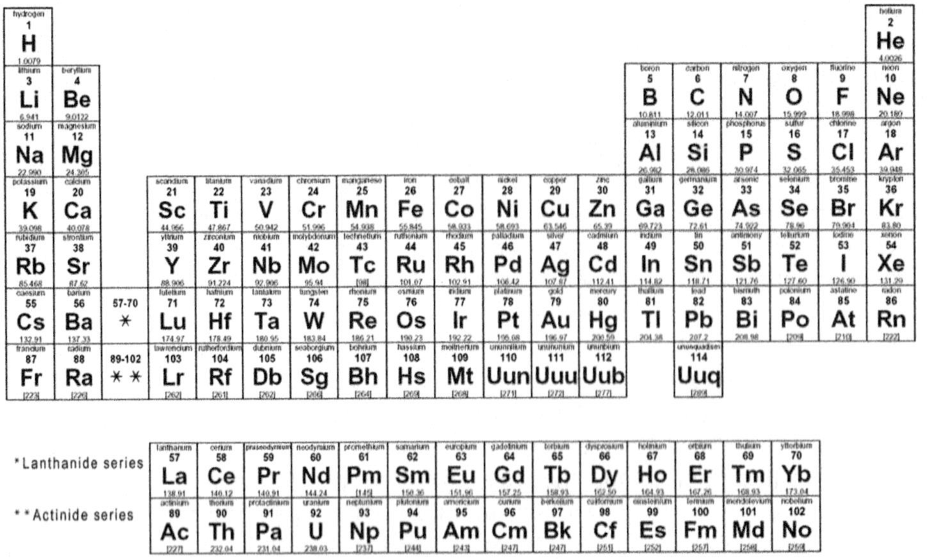

Figure 5.5. Modern periodic table of the elements. The bold number above each element's symbol is the atomic number; the small number underneath the symbol is the atomic weight. Figure from LeVanHan at https://commons.wikimedia.org/wiki/File:Periodic-table.jpg, available under a Creative Commons Attribution-Share Alike license (https://creativecommons.org/licenses/by-sa/3.0/deed.en).

in modern dance[5] required each dancer to use different moves[6]. In quantum mechanics, however, only certain moves are allowed. There are more different moves at higher energy than at lower energy. This makes a certain amount of sense, even in the dance analogy—there are many fewer ways to stand still than to dance with wild abandon.

In the simplest version of the theory, there are two possible ways to move at the lowest possible energy (a mirror-image pair), 8 at the next energy, then 18, then 32, in a well-defined mathematical pattern. The electrons thus occupy 'shells' of each energy. If an atom with enough electrons is in its ground state (see Science summary for chapter 2), then there would be two electrons in the first shell, eight in the second, and so on.

The beginnings of this structure are visible in the table. There are two elements in the first row, at which point the first shell is filled. The second row has eight elements, so that in neon (Ne), the first two shells are filled.

But at that point the pattern deviates from the simplistic theory, which assumes that the energy of the electrons is not influenced by the states of the other electrons. This is worth expanding on using the dance analogy.

---

[5] The dance analogy is a common one, although it takes many different forms. Variations of it have been employed by Eddington, Gamow, and Mayer herself.

[6] For a visual example of this, find a clip of the *Charlie Brown Christmas Special* dance online. Not only is each dancer performing different moves, but there is a pair of twins who differ only in that their pattern is a mirror image of each other, just as is true for pairs of electrons in atoms.

Suppose one of the high-energy moves allowed to you is to circle the center of the dance floor at some large radius, as is often done in a waltz. Another move, normally of equal energy, is to dance toward the center of the dance floor and then back out again, as happens in some square dances.

But now assume someone else is waltzing. If they're waltzing in a smaller circle than your move, you're not prevented from waltzing, because making a bigger circle is counted as a different move. In fact, your waltz is largely unaffected by them. But if you try to square dance, you'll need to cross their path every time you move in toward the center and back out again. It can be done without running in to them, but it takes a little extra effort to coordinate. That means that two moves that would be equal energy if no one else was around, now have slightly different energies because of what someone else is doing.

This effect splits the shells into subshells. In each shell, for example, there are only two waltz type motions allowed, similar to going in a big clockwise circle while twirling clockwise, or going in a big clockwise circle while twirling counterclockwise[7]. Waltz moves are the only ones allowed at the lowest energy; that's hydrogen and helium in the table. At the next energy, waltz moves are allowed (but in a bigger, faster circle), employed by the outermost electrons in lithium and beryllium, but six different 'square dance' moves are also introduced, employed by the outermost electrons in boron through neon. Since these electrons have to worry about the waltzing electrons, they have a slightly higher energy, and thus come later in the table. Sodium and magnesium add yet a third circle of waltzers, and aluminum through argon a second set of more energetic square dancers. At this point, quantum mechanics allows yet a third type of move—maybe something appropriate to a mosh pit. The moshers tend to flail about the dance floor crazily, meaning that they feel the interference from other dance styles quite strongly. In fact, while moshers are technically allowed in the third shell, it is easier to go to a fourth subshell of waltzers rather than to add the moshers into the third shell right away. That fourth subshell of waltzers is utilized by potassium and calcium. Only after they are brought into the dance are the third shell moshers introduced in the elements scandium through zinc. Then the fourth shell square dancers come in for the elements gallium through krypton.

In the heaviest atoms, we need one more type of move, inherently more disruptive even than the moshers; for the purposes of our analogy we'll use ballet. Imagine the difficulty of performing a ballet routine, with its long runs, leaps, and energetic twirls, in the midst of a dance floor occupied by waltzers, square dancers, and moshers!

Thus, the shape of the periodic table is explained. In 1941, Mayer had in fact helped complete the shape of the table, in effect arguing that the actinide group of

---

[7] Since clockwise from above is counterclockwise from below, these are the only distinct possibilities. Going in a counterclockwise circle while twirling clockwise, for instance, is the same motion as going in a clockwise circle while twirling counterclockwise. Of course, this isn't *actually* true when dancing, because the real world is much more complicated than the simple world inside the atom, with dance floors, gravity, walls, etc.

elements seen in figure 5.5 are employing the same type of ballet move as the lanthanide series [15].

Whenever a subshell, or particularly a shell, is filled, it is particularly stable. Thus helium, neon, argon, etc on the right side of the table are 'noble' gasses which rarely undergo chemical reactions.

All of this was well understood by the mid 1940s. There were numbers of electrons that corresponded to particularly stable atoms and ions, most notably the atomic numbers of the noble gasses: 2, 10, 18, 36, 54, and 86.

Could the magic numbers have a similar origin, but this time for neutrons and protons in the nucleus, rather than electrons in the outer regions of the atom?

Mayer realized the case would be strengthened if it relied not only on the stability of certain nuclei, but on the periodic properties of nuclei with non-magic numbers of neutrons and protons, just as the system of the periodic table is justified not only by the stability of the noble gasses, but by the similarity of non-noble elements in the same column, such as copper (Cu), silver (Ag), and gold (Au). This was a focus of the first of her papers from 1950 [45].

But it would also be crucial to identify the subshells involved, and why the order of energies was the way it was.

One big difference between the physics of electrons in an atom and the physics of the nucleus: the nucleus is much more crowded! This led to the 'liquid drop' model discussed in the science summary for chapter 2. In this model, the protons and neutrons are in constant contact, sliding past each other like molecules in a liquid. At first, it was assumed that this would destroy any shell structure of the type seen for electrons in atoms.

The dance analogy can again prove useful. Imagine our dancers have the same library of moves available to them as before, and they still are required to have a different move from anyone else on the floor, but now the floor is tightly packed, so that dancers are bumping into each other nearly constantly. You might be trying to perform a slow waltz, but you are getting bumped into so much, you might think you'd easily change from one kind of move into another. But since quantum mechanics doesn't allow you to use a move someone else is using, there might not be many convenient moves available to switch to. Thus, it is much more likely for a nucleon to stay in the same state after a collision than it would be if that rule didn't apply.

The same basic library of moves is available; two low-energy waltzes, two mid-energy waltzes, two high-energy waltzes, and two ultra-high-energy waltzes; six mid-energy square dance moves, six high-energy square dance moves, and six ultra-high-energy square dance moves; ten high-energy mosher moves and ten ultra-high-energy mosher moves; fourteen ultra-high-energy ballet moves; as well as moves with even higher energy. The interference between dancers is now so high that the original energy order is completely scrambled. There isn't much energy difference anymore between a waltzer and a ballet dancer—both are getting hit all the time. So we expect to find moshers and ballet dancers appearing much earlier in the sequence than we did for the case of electrons in atoms. For that reason, I'll drop the original labels of low, mid, high, and ultra-high energy.

Both Mayer and Jensen attempted to calculate the energy order under the conditions in the nucleus. While they used the same assumptions they both got the same results: under the conditions of the nucleus, the order should be [32]:
- two waltzers
- six square dancers
- ten moshers
- two more waltzers, more energetic than the first pair
- fourteen ballet dancers
- six more square dancers, more energetic than the first set

At first, this goes according to plan. The first magic number is two, corresponding to the first two waltzers. Next comes eight, meaning the first set of square dancers get added it. The next one is 20, which must mean the ten moshers and the two new waltzers come in more or less together. (This is turning into a heck of a party!)

But then we get to 28—how does that happen? Perhaps the 14 ballet dancers are skipped, but that's not what Mayer's and Jensen's separate calculations said should happen. And the above sequence suggests a magic number at 40, but the next one after that is 50. Something was wrong.

This is when Fermi asked about spin–orbit coupling. (In Jensen's case, he came up with that idea on his own, perhaps while he was shaving [54].)

On the packed dance floor, it's possible that all the dancers are jostling in different directions. But it's also possible that, despite doing different moves, there's an overall circulation—that the entire mass of dancers is tending to orbit the floor in the same direction, despite occasional exceptions. If that were the case, it would be much easier to also twirl in the same direction as that overall motion, rather than against it. That's what's meant by spin–orbit coupling.

In the case of electrons in an atom, spin–orbit coupling was known to exist, but it was a very small effect. In the crowded conditions of the nucleus, though, it turned out to be a large effect.

Some combinations of moves would lead to an overall circulation, and some wouldn't. When the number was right so that there was a circulation, and also allowed more individual dancers to twirl with the circulation than against it, then that combination would be lower in energy than would otherwise be expected, and thus be more stable.

Taking in to account spin–orbit coupling, the new order became:
- two waltzers
- four square dancers with favorable twirling
- two square dancers with unfavorable twirling
- six moshers with favorable twirling
- two more waltzers, more energetic than the first pair
- four moshers with unfavorable twirling
- eight ballet dancers with favorable twirling
- four more square dancers, more energetic than the first set, with favorable twirling
- six ballet dancers with unfavorable twirling

- two more square dancers, more energetic than the first set, with unfavorable twirling
- ten of whatever comes after ballet dancers, with favorable twirling

Twenty eight now comes out naturally, with the addition of the first set of ballet dancers, and 50 with everything that is listed above. Of course, the sequence continues, explaining 82 and 126 as well. This is what Mayer worked out in the ten minutes after Fermi asked about spin–orbit coupling. As she put it in her Nobel lecture:

All the magic numbers are explained in the same way. And since they are explained and no longer magic, I shall from here on call them shell numbers [32].

She and Jensen then put strong spin–orbit coupling into their theoretical models, and came up with the sequence of energy levels found above.

If the theory was to be shown to be correct, it should also predict properties, such as the total circulation, of nuclei with numbers of protons and neutrons that were not magic. Mayer demonstrated that in many cases it did, although there were discrepancies [45]. As she said in the conclusion to her Nobel lecture:

The shell model has initiated a large field of research. It has served as the starting point for more refined calculations. There are enough nuclei to investigate so that the shell model lists will not soon be unemployed [32].

Among the research initiated were some of the last scientific papers of Lise Meitner, investigating the relevance of the shell model to nuclear fission [55].

References

[1] Dash J 1973 *A Life of One's Own: Three Gifted Women and the Men They Married* (New York: Harper and Row)
[2] Mayer and Maria Goeppert 2008 *Complete Dictionary of Scientific Biography* http://www.encyclopedia.com/people/science-and-technology/physics-biographies/maria-goeppert-mayer
[3] Sachs R G 1979 Maria Goeppert Mayer *Biographical Memoirs* **50** 310–28
[4] Interview of J Franck and H S Franck by Thomas S Kuhn and Maria Goppert Mayer on July 14, 1962, Niels Bohr Library & Archives, American Institute of Physics, College Park, MD USA, www.aip.org/history-programs/niels-bohr-library/oral-histories/4609-6
[5] Interview of Franck J and Franck H S by Thomas S Kuhn and Maria Goppert Mayer on July 14, 1962, Niels Bohr Library & Archives, American Institute of Physics, College Park, MD USA, www.aip.org/history-programs/niels-bohr-library/oral-histories/4609-5
[6] Des Jardins J 2010 *The Madame Curie Complex: The Hidden History of Women in Science* (New York: Feminist Press)
[7] Lewin Sime R 1996 *Lise Meitner: A Life in Physics* (Berkeley, CA: University of California Press)

[8] Bird K and Sherwin M J 2005 *American Prometheus: The Triumph and Tragedy of J Robert Oppenheimer* (New York: Knopf)
[9] Bonolis L *Maria Goeppert Mayer, Mediateque* accessed April 9 2017, http://www.mediatheque.lindau-nobel.org/research-profile/laureate-goeppert-mayer
[10] Wang Z 2000 Mayer, Maria Gertrude Goeppert *American National Biography Online* February 2000 (last modified: February 18, 2003) http://www.cpp.edu/~zywang/Mayer.htm
[11] Joseph Mayer *Array of Contemporary American Physicists* accessed April 9, 2017. http://history.aip.org/history/acap/biographies/bio.jsp?mayerj
[12] Wright J K and Carter G F 1959 Isaiah Bowman *Biographical Memoirs* **33** 38–64
[13] Mayer P C 2011 *Son of (Entropy)^2* (Bloomington, IN: AuthorHouse)
[14] Mayer J E and Goeppert Mayer M 1940 *Statistical Mechanics* (New York: Wiley)
[15] Goeppert Mayer M 1941 Rare-earth and transuranic elements *Phys. Rev.* **60** 184–7
[16] Nation's Defense Takes Toll of S L Faculty *Campus* (Yonkers, NY) January 14, 1942, 3
[17] Jing Min Chia Maria 2010 *Goeppert Mayer: Revisiting Science at Sarah Lawrence* (unpublished)
[18] Interview of J Mayer by Lillian Hoddeson on January 24, 1975, Niels Bohr Library & Archives, American Institute of Physics, College Park, MD USA, https://www.aip.org/history-programs/niels-bohr-library/oral-histories/4769
[19] Contract, Constance Warren to Maria Goeppert Mayer, June 22, 1942 *Sarah Lawrence College Archives*
[20] Howes R H and Herzenberg C L 1999 *Their Day in the Sun: Women of the Manhattan Project* (Philadelphia: Temple University Press)
[21] Letter, Maria G Mayer to Constance Warren, October 6, 1943, *Sarah Lawrence College Archives*
[22] Letter, Maria Goeppert Mayer to Constance Warren, February 7, 1944 *Sarah Lawrence College Archives*
[23] Contract, Constance Warren to Maria Goeppert Mayer, April 22, 1944 *Sarah Lawrence College Archives*
[24] Letter, Constance Warren to Maria Goeppert Mayer, June 22, 1942, as amended December 22, 1944 *Sarah Lawrence College Archives*
[25] Wartime Trends is Topic at Faculty Round Table; Lowenberg, Roesch, Mayer, and Lynd Speak *Campus* (Yonkers, NY) February 28, 1945, 1
[26] Goeppert Mayer M *The Reminiscences of Maria Goeppert Mayer, Columbia Center for Oral History Collection* quoted in *Maria Goeppert Mayer: Revisiting Science at Sarah Lawrence College, Sarah Lawrence College Archives*, accessed April 9, 2017 https://www.sarahlawrence.edu/archives/exhibits/maria-goeppert-mayer-exhibit
[27] *Sarah Lawrence College Course Catalogues, 1944-1945*, quoted in Jing Min Chia *Maria Goeppert Mayer: Revisiting Science at Sarah Lawrence* (unpublished).
[28] Contract, Constance Warren to Maria Goeppert Mayer, June 6, 1945 *Sarah Lawrence College Archives*
[29] Kimball A H interview by Eleni Digenis, *Interviews of the Margaret MacVicar Memorial AMITA Oral History Project* February 16, 1993
[30] Letter, Maria Goeppert Mayer to Talyor, August 15, 1945 *Sarah Lawrence College Archives*
[31] Letter, Harold Taylor to Maria Goeppert Mayer, August 20, 1945 *Sarah Lawrence College Archives*.

[32] Goeppert Mayer M 1970 The Shell Model *Nobel Lectures Physics 1963-1970* (Amsterdam: Elsevier)
[33] Matthews R 1994 Science: The ghostly hand that spaced the planets *New Scientist* April **9**
[34] Teller E and Schoolery J 2001 *Memoirs: A Twentieth-Century Journey in Science and Politics* (Cambridge, MA: Perseus)
[35] Mayer M G 1948 On closed shells in nuclei *Phys. Rev.* **74** 235–9
[36] Jensen J and Hans D 1970 Glimpses at the history of the nuclear structure theory *Nobel Lectures Physics 1963-1970* (Amsterdam: Elsevier)
[37] Feenberg E and Hammack K C 1949 Nuclear shell structure *Phys. Rev.* **75** 1877–93
[38] Nordheim L W 1949 On spins, moments, and shells in nuclei *Phys. Rev.* **75** 1894–901
[39] Goeppert Mayer M 1949 On closed shells in nuclei. II *Phys. Rev.* **75** 1969–70
[40] Feenberg E, Hammack K C and Nordheim L W 1949 Note on proposed schemes for nuclear shell models *Phys. Rev.* **75** 1968–9
[41] Suess H 1984 interview by Karen Fleckenstein, *UCSD Oral Histories*, October 24
[42] Haxel O, Jensen J, Hans D and Suess H E 1949 On the 'magic numbers' in nuclear structure *Phys. Rev.* **75** 1766
[43] Haxel O, Jensen J, Hans D and Sueß H E 1948 Zur Interpretation der ausgezeichneten Nucleonenzahlen im Bau der Atomkerne *Nature* **35** 376 (note that despite the nominal 1948 publication date, the paper was submitted and published in 1949)
[44] Jensen J, Hans D, Sueß H E and Haxel O 1949 Modellmäßige Deutung der ausgezeichneten Nucleonenzahlen im Kernbau *Nature* **36** 155–6
[45] Goeppert Mayer M 1950 Nuclear configurations in the spin-orbit coupling model. I. empirical evidence *Phys. Rev.* **78** 16–21
[46] Goeppert Mayer M 1950 Nuclear configurations in the spin-orbit coupling model. II. theoretical considerations *Phys. Rev.* **78** 22–3
[47] Goeppert Mayer M and Jensen J Hans D 1955 *Elementary Theory of Nuclear Shell Structure* (New York: Wiley)
[48] Maria Goeppert Mayer —Nominations *Nobel Prizes and Laureates*, accessed April 8, 2017 https://www.nobelprize.org/nobel_prizes/physics/laureates/1963/mayer-nomination.html
[49] Hans J and Jensen D —Nominations *Nobel Prizes and Laureates*, accessed April 8, 2017 https://www.nobelprize.org/nobel_prizes/physics/laureates/1963/jensen-nomination.html
[50] The Nobel Prize in Physics 1963 *Nobel Prizes and Laureates* accessed April 8, 2017 https://www.nobelprize.org/nobel_prizes/physics/laureates/1963
[51] All data in this section based on information at *Nobel Prizes and Laureates*, accessed April 8, 2017 http://www.nobelprize.org/nobel_prizes
[52] Alfred Bernhard Nobel, will, translated as *Full Text of Alfred Nobel's Will* Nobel Media B 2014. Web. 9 Apr 2017 http://www.nobelprize.org/alfred_nobel/will/will-full.html
[53] The Noble Prizes in the new century: An interview with Ralf Pettersson, Director of the Stockholm Branch of the Ludwig Institute for Cancer Research, the Karolinska Institute, and former chairman of the Nobel Prize Committee for Physiology/Medicine *EMBO Reports* **2** (2001) 83–85
[54] Fleckenstein Karen interviewing Hans Suess, *UCSD Oral Histories*, October 24, 1984
[55] Meitner L 1950 *Nature* **165** 561

IOP Concise Physics

Beyond Curie
Four women in physics and their remarkable discoveries, 1903 to 1963
Scott Calvin

Chapter 6

Afterword

One of the great benefits of focusing on four prominent women in physics, rather than just one, is that it mitigates the idea that each one has to stand in for her entire gender, at least within that profession. It becomes clear that Wu's assertive style is not the only one possible for a woman in the field, any more than Mayer's deference is. They are different people, with different strengths and personalities; there are many ways to be a genius.

But are there commonalities between them? Four is still a pretty small sample size, and it also suffers from selection bias: I chose four prominent female physicists who interest me, and make no claim that these four are the most prominent from the period. Still, I think we can see a few common threads.

One is indisputable: each of these scientists was, often for large stretches of her career, underpaid. There seems to have been no way for them to defend against this problem: it happened when they were unknown and when they were famous, in the 1920s and in the 1960s, to the married and unmarried, in America and in Germany.

Another is that each faced a delay in her career, somewhere along the line, that probably would not have happened to a man.

A third commonality is more subtle. Each achieved a measure of fame during her lifetime, both within the scientific community and with the general public. But each has also faded from public consciousness over time, perhaps to a greater extent than men of equal stature. I think that is in part due, ironically, to the mythologizing of particular women in science. In these myths, there is one great woman scientist, Marie Curie. At some later date, there is one additional brilliant woman scientist, sometimes explicitly referred to as the '[insert ethnic or national identity] Madame Curie'. This woman works alone in a field that is entirely made up of men, makes a brilliant discovery, has the credit stolen by a man, and then dies in obscurity.

Like all myths, there are variations in the tellings; sometimes, for example, the end emphasizes recognition late in life or posthumously, at other times, the modern obscurity is emphasized. But the central elements appear again and again.

Also like other myths, they contain some elements of truth, and they tell us something about the world, but they also risk distorting history and devaluing the contributions of 'ordinary' people.

In particular, the version of these myths I've outlined above requires suppressing the contribution of other women. It spoils the simple narrative to mention that Mayer and Wu *both* worked at Columbia on the Manhattan Project, or that one of the chief rivals of the team of Meitner and Hahn was the team of Curie and Joliot. Certainly, each worked in a field dominated by men, but not to the point that other women were absent.

The same thing does not happen with men. While minor figures may disappear from the narratives over time, there is room for more than two male scientists in our popular accounts. In narratives, even casual ones, of the development of quantum mechanics, for example, Planck, Rutherford, Einstein, Schrödinger, Bohr, Heisenberg, Pauli, Dirac and Fermi jostle together for space, each boosting the fame of the next, rather than requiring the exclusion of the others.

There are also commonalities that have to do with the experience of all (or most) physicists, regardless of gender. Each has role models and mentors, moments of doubt and periods of obsession, allies and rivals and priority disputes.

The Nobel Prize, with its quirky rules and complicated history, is terrible as a measure of success. But for the general public, when the Swedish Academy awarded a Nobel Prize in physics to Marie Curie in 1903, it simultaneously established Curie as a genius and the Prize as the criterion by which genius is recognized. The Academy did not see fit to recognize another women in physics until 60 years later, when Maria Mayer won. Nearly another 60 years have passed without a winner; Vera Rubin, thought to be a leading contender, passed away in 2016.

A woman should not have to be 'another Madame Curie' to win a Nobel; every year, men win who are not considered to be another Einstein. But winning a Nobel Prize requires a lot of elements to go right: the scientific investigation itself, recognition by the scientific community, clear distinction from other contenders (particularly those who made similar discoveries), political and interpersonal considerations. If just one of those elements goes awry, the prize could go to someone else. A woman could face extra obstacles having to do with any of those aspects, beyond those faced by male colleagues.

If a woman wins the Nobel in physics this year, it would be nice, but on its own it wouldn't change the pattern. We have never seen a female physics Nobel laureate congratulate another on receiving the prize, because there have never been two alive at the same time. Neither have there been two female Nobel laureates in chemistry alive at the same time—Marie missed seeing Irène win the prize by less than two years.

That is the image I hope to see, and see soon—a group photo of living Nobel laureates in physics, featuring two, three, five women[1]. It is not so difficult to imagine. Look at figure 2.6, or figure 3.1. Women, in the plural, have been a part of scientific communities for as long as there has been science. It is long past time for them to receive the supportive conditions that foster excellence, and to be recognized for excellence when they achieve it.

---

[1] I can hope, of course, for one that features 50% women, but the overwhelming statistics among current laureates means that is still a long way off.

CPSIA information can be obtained
at www.ICGtesting.com
Printed in the USA
BVOW07s1528270817
493174BV00002B/17/P

9 781681 746449